Lecture Notes in Mathematics

2222

More information about this series at http://www.springer.com/series/304

Peter Roquette

The Riemann Hypothesis
in Characteristic p
in Historical Perspective

Peter Roquette
Mathematical Institute
Heidelberg University
Heidelberg, Germany

Photo of Emil Artin, outside his apartment in Hamburg, approx. 1935 on page 13
© Estate of Natascha Artin-Brunswick

Photo of Friedrich Karl Schmidt on page 39
Author: Konrad Jacobs, Source © Archives of the Mathematisches Forschungsinstitut
Oberwolfach

Photo of Helmut Hasse, Hamburg, approx. 1934 on page 55
Peter Roquette's personal photo archive

Photo of Harold Davenport, Cambridge approx. 1932 on page 59
© James H. Davenport

Photo of Max Deuring, 1973 on page 136
Author Konrad Jacobs, Source © Archives of the Mathematisches Forschungsinstitut Oberwolfach

Photo of Andre Weil on page 203
© Sylvie Weil personal archives

ISSN 0075-8434   ISSN 1617-9692   (electronic)
Lecture Notes in Mathematics
ISSN 2193-1771
History of Mathematics Subseries
ISBN 978-3-319-99066-8   ISBN 978-3-319-99067-5   (eBook)
https://doi.org/10.1007/978-3-319-99067-5

Library of Congress Control Number: 2018955713

Mathematics Subject Classification (2010): 01A60, 11R58, 14H05

This Springer imprint is published by the registered company Springer Nature Switzerland AG
The registered company address is: Gewerbestrasse 11, 6330 Cham, Switzerland

Preface

This book is the result of many years' work. I am telling the story of the Riemann hypothesis for function fields, or curves, of characteristic p starting with Artin's thesis in the year 1921, covering Hasse's work in the 1930s on elliptic fields and more, until Weil's final proof in 1948. The main sources are letters which were exchanged among the protagonists during that time which I found in various archives, mostly in the University Library in Göttingen but also at other places in the world. I am trying to show how the ideas formed, and how the proper notions and proofs were found. This is a good illustration, fortunately well documented, of how mathematics develops in general.

Some of the chapters have already been prepublished in the "*Mitteilungen der Mathematischen Gesellschaft in Hamburg.*" But before including them into this book, they have been thoroughly reworked, extended, and polished. I have written this book for mathematicians, but essentially it does not require any special knowledge of particular mathematical fields. I have tried to explain whatever is needed in the particular situation, even if this may seem to be superfluous to the specialist.

Every chapter is written such that it can be read independently of the other chapters. Sometimes this entails a repetition of information which has already been given in another chapter. The chapters are accompanied by "summaries." Perhaps, it may be expedient first to look at these summaries in order to get an overview of what is going on.

The preparation of this book had been supported in part by the *Deutsche Forschungsgemeinschaft* and the *Marga und Kurt Möllgaard Stiftung*. A number of colleagues and friends have shown interest and helped me with their critical comments; it is impossible for me to mention all of them here. Nevertheless, I would like to express my special thanks to Sigrid Böge, Karin Reich, Franz-Viktor Kuhlmann, and Franz Lemmermeyer.

References Most of the letters and documents cited in this book are contained in the *Handschriftenabteilung* of the University Library of Göttingen (except if another source is mentioned), most but not all of them in the Hasse files. The letters from

Hasse to Davenport are contained in the archive of Trinity College, Cambridge. The letters from Hasse to Mordell are contained in the archive of King's College, Cambridge. The letters from Hasse to Fraenkel are in the archive of the Hebrew University, Jerusalem.

I have translated the cited text into English for easier reading. Transcriptions of the full letters (and more) in their original language are available at my homepage:

https://www.mathi.uni-heidelberg.de/~roquette/

Heidelberg, Germany   Peter Roquette

The original version of the book was revised. The correction is available at https://doi.org/10.1007/978-3-319-99067-5_14

Contents

1 Overture 1
  1.1 Why History of Mathematics? 1
  1.2 Artin and the Intervention by Hilbert 2
  1.3 Hasse's Project 4
  1.4 Weil's Contribution 6
  Summary 7

2 Setting the Stage 9
  2.1 Our Terminology 9
  2.2 The Theorem: RHp 11

3 The Beginning: Artin's Thesis 13
  3.1 Quadratic Function Fields 14
      3.1.1 The Arithmetic Part 17
      3.1.2 The Analytic Part 20
  3.2 Artin's Letters to Herglotz 25
      3.2.1 Extension of the Base Field 26
      3.2.2 Complex Multiplication 27
      3.2.3 Birational Transformation 30
  3.3 Hilbert and the Consequences 31
  Summary 31
  3.4 Side Remark: Gauss' Last Entry 32
  Summary 37

4 Building the Foundations 39
  4.1 F.K. Schmidt 39
  4.2 Zeta Function and Riemann-Roch Theorem 43
  4.3 F.K. Schmidt's L-polynomial 47
      4.3.1 Some Comments 48
  4.4 The Functional Equation 49
  4.5 Consequences 51
  Summary 53

5   Enter Hasse   55

6   Diophantine Congruences   59
    6.1   Davenport   59
    6.2   The Challenge   61
    6.3   The Davenport-Hasse Paper   67
        6.3.1   Davenport's Letter and Generalized Fermat Fields   67
        6.3.2   Gauss Sums   70
        6.3.3   Davenport-Hasse Fields   73
        6.3.4   Stickelberger's Theorem   75
    6.4   Exponential Sums   76
    Summary   78

7   Elliptic Function Fields   81
    7.1   The Breakthrough   81
    7.2   The Problem   83
    7.3   Hasse's First Proof: Complex Multiplication   85
    7.4   The Second Proof   90
        7.4.1   Meromorphisms and the Jacobian   92
        7.4.2   The Double Field   96
        7.4.3   Norm Addition Formula   97
        7.4.4   The Frobenius Operator   102
    7.5   Some Comments   104
        7.5.1   Rosati's Anti-automorphism   104
    Summary   105

8   More on Elliptic Fields   107
    8.1   The Hasse Invariant A   107
    8.2   Unramified Cyclic Extensions of Degree p   109
        8.2.1   The Hasse-Witt Matrix   111
    8.3   Group Structure of the Jacobian   112
        8.3.1   Higher Derivations   115
    8.4   The Structure of the Endomorphism Ring   117
        8.4.1   The Supersingular Case   118
        8.4.2   Singular Invariants   121
        8.4.3   Elliptic Subfields   122
        8.4.4   Good Reduction   124
    8.5   Class Field Theory and Complex Multiplication   126
    Summary   128

9   Towards Higher Genus   131
    9.1   Preliminaries   131
    9.2   The Years 1934–1935   134
        9.2.1   Deuring   136
        9.2.2   More Political Problems   139
    9.3   Deuring's Letter to Hasse   141
        9.3.1   Correspondences   143

9.4   Hasse's Letter to Weil 146
9.5   Weil's Reply and Lefschetz's Note 148
9.6   The Workshop on Algebraic Geometry 152
Summary 156

10   A Virtual Proof 157
10.1   The Quadratic Form 157
       10.1.1   The Double Field 157
       10.1.2   The Different 159
10.2   Positivity 166
       10.2.1   Applying the Riemann-Hurwitz Formula 166
       10.2.2   The Discriminant Estimate 168
10.3   The RHp 173
10.4   Some Comments 175
       10.4.1   Frobenius Meromorphism 175
       10.4.2   Differents 176
       10.4.3   Integer Differentials 177
10.5   The Geometric Language 177
Summary 179

11   Intermission 181
11.1   Artin Leaves 181
11.2   The Italian Connection 183
       11.2.1   Severi 183
       11.2.2   The Volta Congress 186
11.3   The French Connection 190
       11.3.1   On Function Fields 190
       11.3.2   The Book 195
       11.3.3   Paris and Strassbourg 198
Summary 202

12   A. Weil 203
12.1   Bonne Nouvelle 204
12.2   The First Note 1940 207
12.3   The Second Note 1941 209
Summary 212

13   Appendix 215
13.1   Bombieri 216

Correction to: The Riemann Hypothesis in Characteristic p in
Historical Perspective E1

References 221

Index 231

Chapter 1
Overture

1.1 Why History of Mathematics?

Mathematics is, on the one hand, a *cumulative science*. Once a mathematical theorem has been proved to be true then it remains true forever: it is added to the stock of mathematical discoveries which has piled up through the centuries and it can be used to proceed still further in our pursuit of knowledge.

On the other hand, the mere proof of validity of a theorem is in general not satisfactory to mathematicians. We also wish to know "why" the theorem is true, we strive to gain a better understanding of the situation than was possible for previous generations. Consequently, although a mathematical theorem never changes its content we can observe, in the history of our science, a continuous change of the *form of its presentation*. Sometimes a result appears to be better understood if it is generalized and freed from unnecessary assumptions, or if it is embedded into a general theory which opens analogies to other fields of mathematics. Also, in order to make further progress possible it is often convenient and sometimes necessary to develop a framework, conceptual and notational, in which the known results become trivial and almost self-evident, at least from a formalistic point of view. So when we look at the history of mathematics we indeed observe changes, not in the nature of mathematical truth, but in the attitude of mathematicians towards it. It may well be that sometimes a new proof is but a response to a current fashion, and sometimes it may be mere fun to derive a result by unconventional means. But mostly the changes in attitude reflect a serious effort towards a better understanding of the mathematical universe.

I believe it is worthwhile to observe such trends in the past and see how they have led to the picture of today's mathematics. Here I am telling the fascinating story of the emergence of the Riemann Hypothesis in characteristic p and of its proof. Initiated by Artin in analogy to algebraic number theory, further developed

© Springer Nature Switzerland AG 2018   1
P. Roquette, *The Riemann Hypothesis in Characteristic* p
in Historical Perspective, Lecture Notes in Mathematics 2222,
https://doi.org/10.1007/978-3-319-99067-5_1

by Hasse and Deuring in the framework of function fields, and later embedded into
the new algebraic geometry by A. Weil, this development exhibits all the features of
mathematical research mentioned above.

Our story covers roughly the years from 1921 to 1948 (with an appendix devoted
to Bombieri's proof in 1976). In this period Hasse and his team did not yet reach
a proof for arbitrary function fields of higher genus. But all the prerequisites of a
proof had been available already in the year 1936. In order to provide evidence for
this I have included a chapter with a short "virtual" proof which, indeed, could have
been given in 1936 by Hasse or Deuring or any one of their team.

This story is a good example of what we often observe in the history of
mathematics[1]: that much effort and time had to be spent by the pioneers to
explore their way into unknown territory. Thereby they paved the way for the
next generations who now can travel comfortably along their smooth tracks. I have
extensively used original letters and documents since this provides a glimpse into the
communication channels of mathematicians of that time. One can see the difficulties
which had to be overcome although nowadays they are not any more considered
difficult at all. This book may be regarded as just a commentary to those letters
which had been exchanged at that time.

I have met in my life most of the people from whose letters I am citing but
I have to admit that in the past I did not ask them about their own opinions and
remembrances of those times. Today I regret that I missed the opportunity to get
to know more about the personal memories of these people. Nevertheless, my
interpretation of their letters and notes may show a certain personal touch in as
much as it could possibly depend on my own impressions of their personalities.

1.2 Artin and the Intervention by Hilbert

Our story starts in Göttingen in the early 1920s. The mathematical scene in
Göttingen had recovered from the difficult years of World War I and the immediate
post-war period. Göttingen was again considered as "the" center of mathematics
in the world. It was to become an attractive place in particular for young mathematicians from Germany and from abroad, who were eager to learn about the latest
developments. The mathematical atmosphere was bristling with new ideas—an ideal
background for progress.

But not all the young people arriving in Göttingen found the atmosphere
congenial to their expectations. Let me cite from a letter written by a young Austrian
postdoc to his academic teacher. The letter is dated 30 November 1921. He had
recently arrived in Göttingen and had already introduced himself to the big shots,
i.e., to Courant, Hilbert, Klein and Landau. (Emmy Noether was not yet considered
an important figure in the Göttingen scene whom any newcomer had to formally

---

[1]And not only of mathematics.

introduce himself—although she was the one who cared personally for the young
people arriving in Göttingen, including our man.) He had been friendly received,
and Hilbert had invited him to give a talk at the Mathematical Society of Göttingen
in order to report on his thesis and further work. The talk had been scheduled for
25 November 1921.

However, in his letter five days later he reports that he is deeply disappointed:

*"... I have now given my talk but with Hilbert I had no luck. Landau and the
other number theorists were quite pleased with it, they said so even during my
talk when Hilbert often interrupted me. But he kept interrupting frequently—
finally I could not speak any more at all—and he said that from the start he
did not even listen since he had the impression that everything was trivial.
But then he changed his mind when I mentioned the said decomposition of
prime numbers.[2] I had to do this out of the proper context since I could
not speak and hence could not present the latest results of my thesis and of
my recent investigations. But anyhow this talk had not been successful and
Hilbert, through his criticism, has killed my enthusiasm for this work. By
the way, in my opinion (and in the opinion of the others) his criticism is not
justified. I do not know your opinion about this but as to myself, the delight
for these results is gone."*

The name of this young man was Emil Artin, he was 23 years of age and wrote
this letter to his Ph.D. advisor Gustav Herglotz in Leipzig. Hilbert's conduct came
as a shock to Artin. In fact, Artin closed his letter with the words:

*"Now please excuse me, Herr Professor, from again having bothered you with
such a long letter but this will probably not happen any more with this subject
since I intend to drop it."*

Artin's letter had 5 pages. We have cited the last page only, the earlier pages
contain a report on certain supplements to his thesis. These pointed in the direction
of the RHp[3] but Artin was not able to present them in his talk, due to Hilbert's
interruptions. (See Sect. 3.2.)

Artin's thesis contained the foundations of the arithmetic theory of quadratic
function fields over a finite base field, in analogy to quadratic number fields. It
included the definition and study of the zeta functions of those fields, with the aim
of applying this to class number formulas, density results etc. Artin had added a list
of about 30 numeric cases where he computed the class numbers and, on the way,
he observed the validity of the Riemann hypothesis in characteristic p for these
examples (see Sect. 2.2).

Young Artin's disappointment about Hilbert's unfounded criticism went quite
deep. Indeed he did what he had announced in his letter to Herglotz, namely he

---

[2]See formulas (3.15b) and (3.16b) in Sect. 3.2.2.

[3]Here and in the following I am using the abbreviation "RHp" for "Riemann Hypothesis in function
fields of characteristic p".

turned to other problems and in his publications never took up the problem of the RHp. Also, he left Göttingen and accepted a position at Hamburg University which then witnessed the rapid rise of Artin to one of the top mathematicians of his time.

To vindicate Hilbert it should be said that some days after the talk he admitted that now he appreciated Artin's work and did not any more consider it trivial. He offered Artin the publication of his new results in the *Mathematische Annalen*. But Artin did not accept the offer.

Artin's thesis appeared in print three years later in the *Mathematische Zeitschrift* [Art24]. It is the only one of all his publications in which the RHp is mentioned. In the *Nachlass* of Artin there was found a print-ready manuscript dated November 1921, containing his subsequent investigations which he had reported to Herglotz. But Artin had never submitted this for publication. See [Art00].

1.3 Hasse's Project

It took quite a while until the RHp was taken up by another mathematician. His name was Helmut Hasse. He had refereed Artin's thesis in the "*Jahrbuch für die Fortschritte der Mathematik*"[4] but had not given any sign that he was particularly interested in the RHp—until he had occasion to talk to Artin in November 1932 when Hasse visited Artin in Hamburg and gave a colloquium talk. At that time, not only Artin but also Hasse was an established mathematician in the top ranks. They had exchanged letters since many years, see [FLR14].

In his Hamburg talk 1932 Hasse spoke about the problem of estimating the number of solutions of diophantine congruences, a problem in which he had recently become interested through his friend Harold Davenport. In the ensuing discussion after the talk Artin pointed out that Hasse's problem was equivalent to the RHp—as a consequence of his (Artin's) unpublished results which he had reported in his letter to Herglotz but never published. Thus Hasse's problem on diophantine congruences mutated into the RHp. With this information at hand he and Davenport succeeded quickly with the proof of RHp for generalized Fermat function fields and for Davenport-Hasse fields.

Parallel to this Hasse also considered the problem of the RHp for elliptic function fields. There he could draw on the classic theory of complex multiplication of elliptic functions which he was familiar with. The date of his success in the elliptic case is documented by a letter of Hasse to Mordell of 6 March 1933 where he reports that he has just obtained the proof. But Hasse's second (final) proof for elliptic fields appeared in print three years later only [Has36c]. Hasse's result in the elliptic case received much attention at that time in the mathematical community. It led to an invitation for a special 1-hour lecture at the next conference of the International Mathematical Union which was scheduled for 1936 at Oslo.

---

[4] Today the reviews in this journal are incorporated in the database of "zbMATH".

Thus Artin, although not any more active in this direction, had contributed essentially to the further development by informing Hasse about his results which he had found in 1921 but Hilbert did not wish to take notice of.

By the way, Hasse like Artin had also not found the mathematical atmosphere in Göttingen congenial to his expectations. He had entered Göttingen University at the end of 1918 when he was 20 years of age. There, in his first semester he attended Hecke's course on algebraic numbers and analytic functions (see [Hec87]). This was quite fascinating as he later recalled. But unfortunately Hecke soon left Göttingen and went to Hamburg. Hasse then became interested in Hensel's theory of p-adic numbers but he was told (by Courant) that in Göttingen this was considered but an unimportant side track and hence not worthwhile to study. So Hasse left Göttingen in 1920 and went to Hensel in Marburg where he wanted to learn more about p-adics. (Thus when Artin arrived in November of 1921 Hasse was not any more in Göttingen, the two would meet first in September of 1922 at the annual meeting of the German Mathematical Society (DMV) in Leipzig.)

In Marburg, Hasse soon got his Ph.D. with a thesis on the Local-Global Principle for quadratic forms. In consequence of Hasse's further work the p-adic numbers became an important, indispensable tool of algebraic number theory—contrary to Courant's prophecy which at that time seems to have been the general opinion in the circle around Hilbert.

After his success in the case of elliptic function fields, Hasse pushed towards a proof of RHp for arbitrary function fields with finite base fields. This turned out to become a larger project since it became necessary first to develop generally the algebraic theory of function fields (over arbitrary base fields) including their Jacobians. At that time this was not yet sufficiently established. From today's geometric viewpoint this is seen as part of the transfer of classical algebraic geometry of curves from characteristic 0 to the case when the base field is of arbitrary characteristic p.

In this story we shall meet the names of quite a number of mathematicians who at least for some time joined Hasse in this work. Some of them are[5]:

- *F. K. Schmidt* (1901–1977) who provided the proper definition of the zeta function of an arbitrary function field over a finite base field. Moreover, he proved the Riemann-Roch Theorem for function fields and showed that this is essentially equivalent with the functional equation of the zeta function [Sch31a].
- *Harold Davenport* (1907–1969) who had introduced Hasse to the problem of counting solutions of diophantine congruences. Moreover, jointly with Hasse he showed that for generalized Fermat function fields and related fields (the so-called Davenport-Hasse fields) the zeros of the zeta function are given explicitly by means of Gauss sums. This solved the RHp for these fields [DH34].

---

[5]Most mathematicians which are mentioned in this book have a biographic article in "Wikipedia" or in "Mac Tutor History of Mathematics Archive" or in other openly accessible places; hence I believe it is not necessary here to always include biographical information—except in a few cases when some such information may be of interest in the present context.

- *Ernst Witt* (1911–1991) who gave the first proof of the functional equation for the *L*-series of function fields over a finite base field. (But he never published it.) Moreover, among quite a number of other important contributions, he determined jointly with Hasse the structure of the maximal unramified abelian extension of exponent p (the characteristic of the field) by means of the so-called Hasse-Witt matrix [HW36].
- *Max Deuring* (1907–1984) who provided an algebraic theory of correspondences and so constructed the endomorphism ring of the Jacobian of a function field of arbitrary genus [Deu37], generalizing what Hasse had done in the elliptic case. Moreover he continued Hasse's work on elliptic function fields by determining completely the structure of their endomorphism rings.
- *André Weil* (1906–1998) who had shown interest in Hasse's project right from the beginning. He had visited Hasse in Marburg in the summer of 1933 and they exchanged letters thereafter. With the outbreak of World War II their contact broke down. Weil could escape from the Nazi terror from France to USA. There in the year 1941 he announced a proof of the Riemann hypothesis for arbitrary function fields with finite base field [Wei41].

Thus Weil was able to complete Hasse's project 10 years after it had been started.

1.4 Weil's Contribution

There is a letter of Weil to Artin dated 10 July 1942 in which he informed Artin about the main ideas and some details of his proof. At that time both Weil and Artin resided as refugee immigrants in the USA. This letter, which Weil has included and commented in his Collected Works [Wei80], is more detailed than his announcement in 1941. The fact that the letter was sent to Artin shows that Artin still was regarded as the ultimate expert for the RHp—although he had nothing published on this subject except what was contained in his thesis more than 20 years ago. The final version of Weil's proof appeared 6 years later [Wei48a].

But Weil did much more. In his proof he stressed the analogy of RHp to problems which had been treated within the framework of classical Italian algebraic geometry. But there, algebraic geometry was studied over the complex base field $\mathbb{C}$ only, i.e., in characteristic 0. So Weil had first to make sure that those results which he needed remain valid in characteristic $p > 0$. To this end he wrote a book containing a complete foundation of algebraic geometry in arbitrary characteristic [Wei46]. This was a formidable task. Weil did not only establish the theory of curves in characteristic p (this would have been sufficient for the RHp) but he covered varieties of arbitrary dimension. This enabled him to develop the theory of abelian varieties over any characteristic. In particular his treatment included Jacobian varieties which led to another proof of the RHp [Wei48b].

Weil's "Foundations of Algebraic Geometry" [Wei46] had an enormous influence. In the course of time it led to what today is generally accepted under the name of "arithmetic geometry", i.e., to the use of the language of algebraic geometry in algebra and number theory, and in other branches of mathematics.

After Weil's success there appeared several papers putting into evidence that the RHp could also have been proved within the theory of algebraic function fields, in which it had been started by Artin in 1921 and continued by Hasse. The last(?) word in this endeavor was given by Bombieri in 1972 [Bom74]. He perfected Hasse's idea to search for a proof which works in the function field directly. He concentrated his proof on the Frobenius map, without caring for the general theory of correspondences of curves. Still he relied heavily on Artin's unpublished results of 1921.

Summary

After having received his degree in Leipzig, Artin spent a year 1921/22 as post-doc in Göttingen. In several letters from there to his academic teacher Herglotz, Artin developed some important further results for zeta functions of quadratic function fields beyond his thesis. But these results were never published. The reason was that when Artin reported about it in Göttingen in the presence of Hilbert, the latter critized his work heavily. Although Hilbert later changed his mind and offered Artin publication of his new results, Artin did not accept. He left Göttingen and went to Hamburg where he turned to other problems. He continued to be interested in the RHp but did not publish anything more in this direction.

Ten years later the RHp was taken up by Hasse. In the year 1932 he gave a colloquium talk in Hamburg about the problem of estimating the number of solutions of diophantine equations, a problem in which he had recently become interested through his friend Harold Davenport. In the discussion with Artin the latter pointed out that the Hasse-Davenport problem was equivalent to the RHp for the function fields in question—as a consequence of Artin's unpublished results which he had reported in his letters to Herglotz but never published. Hasse, stimulated by results of Davenport, succeeded quickly with the proof of RHp for what today are called the generalized Fermat function fields and for related fields. Parallel to this Hasse was able to prove the RHp for elliptic function fields. There he could draw on the classic theory of complex multiplication of elliptic functions which he was familiar with. While preparing his proof for publication Hasse found that it is possible to develop much of "complex multiplication" directly in the case of characteristic $p > 0$. He decided to write a new proof on this basis. This appeared in print three years later, in the year 1936. Although it was still limited to the elliptic case, it received much attention at that time in the mathematical community. It led to an invitation for a special 1-hour lecture at the next conference of the International Mathematical Union at Oslo in the year 1936.

After his success in the elliptic case Hasse and his collaborators started a project towards a proof of RHp for function fields of arbitrary genus. A number of new results were reached in this direction, in particular by Deuring. Today we can see that these results were well sufficient to compose a proof of the RHp. However this goal was not reached at the time.

André Weil had shown interest in Hasse's project right from the beginning. During the 1930s they exchanged a number of letters, among others about the envisaged proof of the RHp for function fields of higher genus. Weil discovered that there is a close connection of the problem to results of Severi in classical algebraic geometry. He managed to translate the foundations of algebraic geometry to characteristic $p > 0$ and on this basis solve the RHp for general function fields in the year 1941. His final proof appeared 1948.

Chapter 2
Setting the Stage

2.1 Our Terminology

I have written this book for mathematicians but I do not assume that the reader is
familiar with the basic notions and terminology of the theory of algebraic function
fields in the 1930s. The present section provides an introduction to this. At the same
time I would like to fix the notations which I will be using in this book.

As said above already a global field of characteristic $p > 0$ can be regarded as a
function field with a finite base field.

I am using the term *"function field"* to denote a finitely generated field extension
$F|K$ of transcendence degree 1. Thus a function field is actually not a field but a
field extension. This somewhat queer terminology is historically sanctioned from
the time when the base field K used to be the field of all complex numbers and
hence needed not to be mentioned explicitly. If $x \in F$ is transcendental over the
base field K then F is a finite algebraic extension of the field $K(x)$ of rational
functions over K. The analogy to number fields is apparent: The rational function
field $K(x)$ is considered to be an analogue to the rational number field $\mathbb{Q}$, and then
the finite algebraic extensions of $K(x)$ become the analogues to the finite algebraic
extensions of $\mathbb{Q}$.

In algebraic geometry one considers also finitely generated field extensions of
arbitrary finite degree of transcendency, i.e., function fields "of several variables".
But in this paper a "function field" $F|K$ is always assumed to be of transcendence
degree 1 (except if explicitly said otherwise).

For simplicity I shall always assume that the base field K is relatively algebraically closed within F (except if explicitly said otherwise). Consequently, if F is
a global field of characteristic $p > 0$ then its base field K consists of all elements in
F which are algebraic over the prime field $\mathbb{F}_p$. Hence in this case the base field K is

© Springer Nature Switzerland AG 2018
P. Roquette, *The Riemann Hypothesis in Characteristic* p
in Historical Perspective, Lecture Notes in Mathematics 2222,
https://doi.org/10.1007/978-3-319-99067-5_2

uniquely determined by F. In general, however, this is not the case. A field F may contain different subfields K, K' such that both $F|K$ and $F|K'$ are function fields in the sense defined above.

Let $F|K$ be a function field. The *prime divisors* of $F|K$ (or simply "primes") correspond to those valuations of F which are trivial on the base field K. If $f \in F$ then its residue modulo a prime P is denoted by $f(P)$, or briefly fP, with the specification that $fP = \infty$ if f does not belong to the valuation ring $\mathcal{O}_P$ of P. Since P is trivial on K its residue field $F(P)$, or briefly FP, contains an isomorphic image of K. Usually the residue map will be normalized such that it leaves K elementwise fixed. The *degree* of P is defined as

$$\deg P = [FP : K].$$

The set of primes P of $F|K$ can be considered as the analogue of what in analysis is called the *Riemann surface* of the function field. Every element $f \in F$ yields a function $f : P \to fP$ on this space; this justifies the name "function field". The value of this function at P is contained in the residue field which is denoted by FP, or we have $fP = \infty$. If $c \in K$ then $cP = c$ for all P; this is the reason why the elements c in the base field K are also called the "constants" of the function field.

Let v_P denote the valuation of $F|K$ belonging to the prime P. Usually v_P is normalized such that the value group $v_P(F^\times) = \mathbb{Z}$ where $F^\times$ denotes the multiplicative group of the field F. If $f \neq 0$ and $fP = 0$ then P is a "zero" of f; in this case $v_P(f) > 0$, and $v_P(f)$ is the "multiplicity" of P as a zero of f. Similarly if $fP = \infty$, then P is a "pole" of f; in this case $v_P(f) < 0$ and $-v_P(f)$ is the multiplicity of the pole. For every $f \neq 0$ we have the relation

$$\sum_P \deg P \cdot v_P(f) = 0, \qquad (2.1)$$

showing that f has as many zeros as it has poles if these are properly counted according to their multiplicities. If K is algebraically closed then every prime is of degree 1 and hence $fP \in K \cup \infty$ for all primes P.

A *divisor* A of $F|K$ is defined to be a formal product of finitely many primes P with positive or negative exponents. The exponent with which P appears in A is denoted by $v_P(A)$. The degree of A is defined by linearity, i.e.,

$$\deg A = \sum_P \deg P \cdot v_P(A).$$

The divisor A is called *principal* if there exists $0 \neq f \in F$ such that $v_P(A) = v_P(f)$ for all P; if this is the case then we write $A = (f)$ or $f \cong A$. The formula (2.1) shows that principal divisors are of degree 0. If AB^{-1} is principal then the divisors A and B are said to be "equivalent" and we write $A \sim B$. Divisors of the same equivalence class have the same degree. Therefore we can speak of the degree of a divisor class C, notation: $\deg C$.

A divisor A is called "integer" if $v_P(A) \geq 0$ for all primes P. If AB^{-1} is integer then the divisor B is said to "divide" A, and A is a "multiple" of B. We write $B|A$ in this situation.

REMARK In our time the language of algebraic geometry is prevalent. In this language a "function field" $F|K$ as defined above is understood as the field of rational functions of an irreducible algebraic curve Γ defined over K, and one writes $F = K(\Gamma)$. If Γ is smooth and complete then it is uniquely determined by the function field $F|K$ (up to birational equivalence over K) and the primes P of $F|K$ are the K-closed points of Γ. In this book I mostly use the language of function fields which stresses the analogy to number fields, and which was mainly used at the time I am reporting about. The translation into the language of geometry is straightforward and easy. (But this was not the case in the 1930s when algebraic geometry was not yet completely included in the "modern" algebra of the time, although a number of attempts in this direction can be registered.) Note that in the framework of algebraic geometry, divisors are usually written in additive notation; then a divisor will be not a formal product but a *formal sum* of primes with positive or negative integer coefficients. Accordingly, an integer divisor A is then called "positive" and one writes $A \geq 0$ in this case.

2.2 The Theorem: RHp

In this book I will tell the story of the Riemann hypothesis in characteristic p. Let me explain:

Let F be a global field of characteristic $p > 0$, i.e., a function field over a finite base field. The zeta function of F is defined by

$$\zeta_F(s) = \prod_P \frac{1}{1 - |P|^{-s}} = \sum_A |A|^{-s} \qquad (2.2)$$

where P ranges over the primes of F and A over the integer divisors of F. The symbol $|P|$ denotes the number of elements in the respective residue field. The integer divisors A of F are formal products of the primes, and $|A|$ is defined multiplicatively. In formula (2.2), s denotes a complex variable. The infinite sum and product converge if the real part $\Re(s) > 1$, but the function $s \to \zeta_F(s)$ extends uniquely to a meromorphic function defined on the whole complex plane, periodic with the period $\frac{2\pi i}{\log q}$ and poles of order 1 at $s = 1$ and $s = 0$.

Riemann Hypothesis in characteristic p. *All zeros of $\zeta_F(s)$ are situated on the line with real part $\Re(s) = \frac{1}{2}$.*

Of course, Riemann himself did not consider fields of characteristic $p > 0$ and their zeta functions. So I should better speak of *the analogue* of the Riemann hypothesis.

But for brevity I will mostly omit the word "analogue" if no misunderstanding seems possible. I shall often use the abbreviation

RHp   for   *Riemann hypothesis in characteristic p* .

Actually, the analogy between the characteristic 0 case and the characteristic p case is not quite straightforward. For, if F is a global field of characteristic 0, i.e., a number field of finite degree, then the corresponding zeta function $\zeta_F(s)$ is defined, due to Dedekind, by the same formula (2.2), where now P ranges over the *prime ideals* of F and A over the *integer ideals* of F. If $F = \mathbb{Q}$ then this is the classical zeta function of Riemann. But now, in characteristic 0, this function does have zeros which are situated on the negative real axis; these are called "trivial". The classical hypothesis of Riemann asserts that the *non-trivial* roots of the zeta function of a number field have real part $\Re(s) = \frac{1}{2}$. The existence of the trivial zeros is due to the fact that in characteristic 0 the product in (2.2) does not contain any term for the infinite primes, which belong to the archimedean valuations of the field. If the infinite primes, represented by certain Gamma-factors, are included in the defining product for the zeta function then there are no trivial zeros any more and the analogy to the characteristic p case is more perfect, at least in appearance.

But the structure of the zeta function in characteristic 0 is much more complicated than in characteristic $p > 0$. While the characteristic 0 case of the Riemann hypothesis remains still among the great unsolved problems of mathematics, the case of characteristic $p > 0$ has been settled. This happened during the decades from the 1920s to the 1950s, and it is the story narrated in this book.

Chapter 3
The Beginning: Artin's Thesis

Emil Artin (1898–1962) was born in Vienna. He was brought up in Reichenberg, a German speaking town in Northern Bohemia belonging to the Austro-Hungarian empire. (The town is now called Liberec, in the Czech Republic). In 1916 he enrolled at the University of Vienna where, among others, he attended a lecture course by Ph. Furtwängler. After one semester of study he was drafted to the army. In January 1919 he entered the University of Leipzig. I have taken this information from Artin's own hand-written vita that he submitted together with his thesis to the Faculty at Leipzig University.[1] There in June 1921 he obtained his Ph.D. with Herglotz as his thesis advisor.

---

[1] The documents for Artin's Ph.D. are kept in the archive of the University of Leipzig.

© Springer Nature Switzerland AG 2018
P. Roquette, *The Riemann Hypothesis in Characteristic* p
in Historical Perspective, Lecture Notes in Mathematics 2222,
https://doi.org/10.1007/978-3-319-99067-5_3

Gustav Herglotz (1881–1953), also of Austro-Bohemian origin, was an all-round mathematician whose work covered astronomy, mathematical physics, geometry, applied mathematics, differential equations, potential theory, and also number theory. His five number theoretical papers were published within the period of 1921–1923. It seems that Artin came to Leipzig just during the time when Herglotz was interested in number theory, and so he inherited this interest from his academic teacher. Or was it the other way round, that Herglotz got interested in number theory through his brilliant student Artin?

In the preface of Artin's Collected Papers [Art65] the editors remark that Artin kept a heartfelt appreciation towards Herglotz throughout his life. Herglotz was the only person whom Artin recognized as his academic teacher. (That's what he answered when once I had asked him about it). Ullrich [Ull00] observes that in Artin's early letters to Herglotz in 1921/22, he always signed with the words "*your grateful disciple*" (*Ihr dankbarer Schüler*).

3.1   Quadratic Function Fields

In his thesis [Art24] Artin considers quadratic function fields, i.e., quadratic extensions of the rational function field $K(x)$. The base field K is assumed to be finite of characteristic $p > 2$. Thus a quadratic function field is of the form $F = K(x, \sqrt{D})$ where $D \in K[x]$ is square free. D is called the "discriminant". Such fields are called "hyperelliptic". But strictly speaking, this applies only if $\deg D > 4$. If $\deg D = 3$ or 4 then F is "elliptic". If $\deg D = 1$ or 2 then F is rational. In order to avoid discussion of trivial cases, let us assume here that $\deg D > 2$ although Artin in his discussion carries the case $\deg D = 0, 1, 2$ as far as possible.

REMARK The case of characteristic 2 can be discussed in a similar manner by using the Artin-Schreier generators of quadratic field extensions. But the theory of Artin-Schreier extensions in characteristic p did not yet exist at the time of Artin's thesis; it was published in 1927 only [AS27b]. The arithmetic of these Artin-Schreier extensions in characteristic p has been developed by Hasse in the year 1934 [Has34e].

Actually, Artin in his published thesis works over the prime field $K = \mathbb{F}_p$ only. But in a letter to Herglotz dated 13 November 1921 he says:

"*It should be observed that the theory remains valid word by word over an arbitrary Galois field, if p is understood to be not a prime number but the corresponding prime power where the exponent is not important. Of course this is self-evident and not new.*"

This letter was written one month after Artin had sent his manuscript to the *Mathematische Zeitschrift*, which was on 14 October 1921. Thus very probably Artin knew about this generalization before sending. Then why did he not include this more general case in his manuscript? Perhaps his above letter contains the answer: Artin did not regard this as necessary because the said generalization was self-evident to him.

Accordingly I will discuss Artin's thesis in his spirit, i.e., as if it would refer to an arbitrary finite base field K, of characteristic $p > 2$. I denote the number of elements of K by q. This is a power of p where the exponent, as Artin said in his letter, may be arbitrary. Sometimes in the literature one writes $K = \mathbb{F}_q$.

Artin's aim is to develop the theory of quadratic function fields, including their zeta functions, in complete analogy to the theory of quadratic number fields. It was Herglotz who had suggested this topic to Artin; this has been stated by Artin himself in his vita which he had to submit to the Faculty in Leipzig. Frei [Fre04] forwards the opinion that Herglotz had read a paper by Kornblum [Kor19] that deals with L-series in rational function fields, and therefore he had proposed to Artin to investigate the L-series belonging to quadratic function fields in more detail. This opinion is confirmed by Herglotz's report on Artin's thesis to the Faculty in Leipzig: It starts with a brief description of the results of Kornblum's paper.

The RHp was *not* the main theme of Artin's thesis. In fact, Herglotz says in his report:

> *"The author develops the complete "number theory" in quadratic function fields—to the same extent as it is known today for quadratic number fields."*

And then Herglotz proceeds to state the main points of the arithmetic of quadratic number fields, which Artin had transferred to the function field case:

– foundation of ideal theory,
– theory of units,
– the zeta function and its functional equation,
– number of ideal classes,
– existence of the genera.

And only in a side remark Herglotz mentions that Artin had obtained certain evidence of the curious fact that the non-trivial roots of the zeta function have real part $\frac{1}{2}$.

Artin assumes the theory of quadratic number fields to be well known, but he does not cite any literature for this. When and where did Artin himself learn about quadratic fields? It seems probable that he had learned it in Herglotz's lectures. For, in the summer semester 1919 Herglotz had announced a lecture course "Elementary Number Theory", and in the following semester "Number Theory (Quadratic Number Fields)". And in the summer semester 1920 Herglotz offered three courses, "Algebraic equations", "Geometry of numbers", and "Problem sessions on number theory", altogether 9 hours weekly.[2] Perhaps this amassment of algebraic and number theoretic courses in Leipzig was done to satisfy the thirst for knowledge

---

[2]I am obliged to Frau Dr. Peter from the University of Leipzig for sending me the lecture announcements of the years 1918–1921.

of his eager student Artin? Herglotz's lectures were generally regarded as "pieces of art" (*Kunstwerke*) according to Constance Reid [Rei76].

A short, 9-page preview of Artin's thesis was published 1921 in the *Jahrbuch* of the Philosophical Faculty of Leipzig University [Art21]. There, Artin says that his proof of the finiteness of class number is analogous to the proof as presented in Weber's Algebra book [Web08]. So we may suppose that Artin had read Weber, as had probably every young German speaking (or at least German reading) mathematician at that time who was interested in algebraic number theory. At one point in his thesis he refers in some detail to Landau's book "Introduction to the elementary and analytic theory of algebraic numbers and ideals" [Lan18].

On the function field side, Artin refers to Dedekind's paper [Ded57]. There, Dedekind had developed the arithmetic theory of the polynomial ring $\mathbb{F}_p[x]$ over the prime field $\mathbb{F}_p$, including the quadratic reciprocity law in this ring—in complete analogy to the arithmetic of $\mathbb{Z}$ and Gauss' quadratic reciprocity law there. Artin says that this suggests to extend Dedekind's theory by adjoining a quadratic irrationality $\sqrt{D(x)}$ to the field of rational functions modulo p. Thus he wants his work to be considered as a continuation of Dedekind's paper which had appeared more than 65 years before. The title of that old paper of Dedekind reads:

"Outline of a theory of higher congruences with respect to a real prime number modulus."

This title may sound somewhat strange to us, but it makes sense if we recall that in those times a "real prime number" was understood to be an ordinary prime number $p \in \mathbb{Z}$, in contrast to "imaginary" primes which may be primes in $\mathbb{Z}[\sqrt{-1}]$ or in other rings of algebraic numbers. And "higher congruences" in $\mathbb{Z}[x]$ meant simultaneous congruences modulo p and modulo some polynomials $f(x) \in \mathbb{Z}[x]$; this is essentially equivalent to congruences in $\mathbb{F}_p[x]$ modulo some polynomials $\overline{f}(x) \in \mathbb{F}_p[x]$. Dedekind regarded $\mathbb{F}_p[x]$ not as a mathematical structure in its own right, but as the result of reduction of $\mathbb{Z}[x]$ modulo p.

On first sight Artin seems to adopt the same viewpoint since the title of his thesis reads:

"Quadratic fields in the domain of higher congruences."

Since Artin uses Dedekind's terminology of "higher congruences", one may be tempted to conclude that he too wants to regard his fields as obtained by reduction mod p from a function field in characteristic 0. But in the very beginning of his paper he says: *"We will call functions and numbers to be equal if they are congruent modulo p in the sense of Dedekind."* This makes clear that, although Artin wishes his work to be regarded as a follow up of Dedekind's, he immediately switches to the viewpoint, "modern" at the time, that he is working in fields of characteristic p in the sense of Steinitz' great paper [Ste10].

REMARK For some time function fields over finite base fields were called "congruence function fields" (*Kongruenzfunktionenkörper*) and their zeta functions were called "congruence zeta functions" (*Kongruenzzetafunktionen*); see, e.g.,

[Has34c, Roq53, WZ91]. This terminology can be understood as a remnant of Dedekind's "higher congruences".

Artin's thesis is divided into two parts:

I. Arithmetic Part.
II. Analytic Part.

The adjectives "arithmetic" and "analytic" are Artin's. But what is their meaning in this context?

Artin does not give any explanation; obviously he assumes that the reader will know how these words were used at the time. But the usage of the word "arithmetic" has changed through the times, and even at the time of Artin's thesis it was not used uniformly. Emmy Noether, for instance, used "arithmetic" in connection with anything referring to ideal theory and prime ideals, and in an extended way also to module theory. Today this would be regarded as belonging to "algebra", or sometimes to "algebraic geometry" but not to "arithmetics". Some authors, also in Artin's time, would reserve the word "arithmetic" for those topics that refer explicitly to properties of the natural numbers. But do function fields in characteristic p fit into this scheme? Similar comments can be given for the use of "analytic". If this should imply that analytic functions are involved then one could point out that the zeta function of a function field is essentially a rational function. The Riemann hypothesis refers to the zeros of a polynomial. Is this "analytic" or "algebraic"?

So let us refrain from the attempt to describe in detail the meaning of "arithmetic" versus "analytic" in the context of Artin's thesis. Artin himself seems to have used these adjectives in some intuitive way, hoping that the reader will be able to follow his intentions; let us try to do this too. Within the theory of function fields, "analytic" will be used for those topics which refer to the zeta function, L-series and similar objects of function fields. Whereas "arithmetic" is used for those topics which do not rely on the use of analytic functions. The analogy to algebraic number theory is prevalent.

3.1.1   The Arithmetic Part

As said above, K denotes a finite field of characteristic $p > 2$. Let $F = K(x, \sqrt{D})$ where $D \in K[x]$ is assumed to be square free. Let $R = K[x, \sqrt{D}]$ denote the integral closure of $K[x]$ in F. (The notation is ours, not Artin's. Quite generally, for the convenience of the reader I use consistently my own notation, which often is different from the various notations used in the papers which I will report on).

In his first part Artin sets out to develop the "arithmetics" of the ring R. His main results are the following statements which we have already seen mentioned in Herglotz's report (see page 15).

(i)   **Prime ideal decomposition:** *R is a Dedekind ring.*
(ii) **Ramification:** *Let $P \in K[x]$ be a prime polynomial. P is ramified in R if and only if P divides D.*
(iii) **Unit Theorem:** *The unit group $R^\times$ is either a torsion group, consisting of the constants, or there exists a fundamental unit ε which generates $R^\times$ modulo torsion—according to whether x has only one pole or two different poles in F.*
(iv) **Class number:** *The ideal class group of R is finite.*
(v)   **Decomposition law:** *Suppose the prime polynomial $P \in K[x]$ is relatively prime to D. Then P splits in R into two different prime ideals if and only if D is a quadratic residue modulo P. Otherwise P remains prime in R.*
(vi) **Reciprocity:** *In $K[x]$ a quadratic reciprocity law holds, in analogy to Gauss' quadratic reciprocity law in $\mathbb{Z}$.*

Artin does not use the terminology of "Dedekind ring"; this was introduced much later. (Perhaps "Bourbaki" can be regarded as the one who created this terminology?) Artin just proves that every proper ideal of R admits a unique representation as a product of prime ideals. In fact, he does not even present a complete proof. After having developed the relevant facts about ideals and their norms in R he says: "*It is seen that the arguments are completely parallel to those in the number field case. Hence in the following it will suffice to state the definitions and theorems.*" I have said earlier already that Artin does not cite any source where the reader could find in detail those arguments for number fields which he uses for function fields. Obviously he could assume that the material was common knowledge among the prospective readers of his thesis.

It is worthwhile to observe how Artin handles the place at infinity. He realizes that in $K(x)$ the place $x \mapsto \infty$ plays a role similar to the ordinary absolute value in the number field case. Now in the latter case, the completion of $\mathbb{Q}$ with respect to the ordinary absolute value is $\mathbb{R}$, the field of real numbers. The theory of quadratic number fields $\mathbb{Q}(\sqrt{D})$ looks quite different according to whether $\sqrt{D}$ is real or not. Here, "real" means that D is a square in $\mathbb{R}$, and then $\mathbb{Q}(\sqrt{D})$ can be regarded as a subfield of $\mathbb{R}$.

In analogy to this, in the function field case Artin considers the valuation of $K(x)$ at infinity and its completion. He does not use the terminology of valuation or completion, he just says that he will extend $K[x]$ by considering not only polynomials but arbitrary power series of the form $f(x) = \sum_{v=-\infty}^{n} a_v x^v$ with $a_v \in K$, which is to say Laurent series with respect to x^{-1}. If $a_n \neq 0$ then Artin defines n to be the degree and $|f(x)| = q^n$ as the size of that Laurent series. (q denotes the number of elements of K). In fact, this is the valuation at infinity. And the field of those Laurent series is the completion of $K(x)$ with respect to this valuation.

The use of this valuation in the function field case is quite the same as that of the ordinary archimedean valuation in the number field case. It is used to estimate ideals and functions in order to prove the finiteness of the class number, and it is also used to define continued fraction expansions in order to obtain the fundamental unit in the "real" case. Here, Artin says that $\sqrt{D}$ is "real" if it can be represented by a Laurent series of the above type, and "imaginary" if not. Artin shows that $\sqrt{D}$ is real if and only if the degree of D is even and, in addition, the highest coefficient of D is a square in K. Of course this is a consequence of Hensel's Lemma but Artin gives explicitly the Laurent expansion of $\sqrt{D}$ in terms of the binomial expansion for the exponent $\frac{1}{2}$.

When Artin wrote his thesis he was not yet aware of the methods and terminology of valuation theory which he later handled so brilliantly. Artin learned the use of valuation theory from Hasse in the years 1923 and later, during the time when Hasse stayed at the University of Kiel not far from Hamburg. They both met frequently in Hamburg at the Hecke seminar. In a letter of July 1923 to Hasse, Artin wrote: *"I am slowly making progress in ℓ-adics. Now I am already taking logarithms!"* (See [FLR14], p. 57).

Concerning the quadratic reciprocity law (vi): If the prime polynomial P does not divide D then Artin introduces the Legendre symbol $(\frac{D}{P})$ which assumes the value 1 or -1 according to whether D is a quadratic residue modulo P or not. Artin points out that Dedekind [Ded57] had stated the quadratic reciprocity law for this symbol without proof, and therefore he will now present a proof in detail. Well, Dedekind had said the following, after stating the reciprocity law: That he had transferred to the function field case all ingredients of Gauss' fifth proof, and therefore it would not be necessary to repeat the proof in every detail.

It seems that this did not satisfy Artin. So he presented his own proof of the quadratic reciprocity law:

$$\left(\frac{Q}{P}\right)\left(\frac{P}{Q}\right) = (-1)^{nm\frac{q-1}{2}} \tag{3.1}$$

where P, Q are different monic prime polynomials in $K[x]$, of degrees n and m respectively. Artin's proof does not follow the lines of Gauss' fifth proof. Frei points out in [Fre04] that Artin's proof can be regarded as the analogue to a proof of Kummer [Kum87].

Apparently Artin did not realize that the quadratic reciprocity law in $K[x]$ for finite K had been proved already by Kühne [Küh02] who more generally dealt also with the power reciprocity law for an arbitrary power ℓ which divides $q - 1$. Probably Artin did not know Kühne's paper.[3] Later, F.K. Schmidt in [Sch27] rediscovered Kühne's power reciprocity law and provided a quite elementary proof. Namely, if $P \neq Q$ are monic irreducible polynomials in $K[x]$ and if

$$P(x) = \prod_{1 \le i \le m} (x - a_i), \qquad Q(x) = \prod_{1 \le j \le n} (x - b_j)$$

---

[3] I am indebted to Franz Lemmermeyer for pointing out Kühne's results to me. Kühne (1867–1907) had obtained his Ph.D. in the year 1892 at the University of Berlin. He was a teacher at the technical school in Dortmund. It would be desirable to obtain more biographic information.

in the algebraic closure of K, then

$$\left(\frac{Q}{P}\right)_\ell = \prod_{i,j}(a_i - b_j)^{\frac{q-1}{\ell}}$$

which trivially gives (3.1) for arbitrary $\ell \mid q - 1$, not only for $\ell = 2$. Here, $(\frac{Q}{P})_\ell$ is the ℓ-the power residue symbol. It is an ℓ-th root of unity, and it is 1 if and only if Q is an ℓ-th power mod P.

Having established the quadratic reciprocity law in $K[x]$ Artin follows straight-forwardly the analogy to quadratic number fields. He defines the Jacobi symbol $(\frac{M}{N})$ for two *arbitrary* polynomials M, N in $K[x]$ without common divisor, in such a way that it becomes bi-multiplicative in the two variables M and N. Formula (3.1) remains valid for this extended symbol if M and N are monic. Artin points out its application to the decomposition law in statement (v) above. Namely, (v) shows that the decomposition type of a monic prime polynomial P, not dividing D, is governed by the value

$$\chi_D(P) = \left(\frac{D}{P}\right) \tag{3.2}$$

which is either 1 or -1. On the other hand, $\chi_D(M) = \left(\frac{D}{M}\right)$ is defined as a Jacobi symbol for an *arbitrary* polynomial $M \in K[x]$ relatively prime to D. It follows from the reciprocity law that $\chi_D(M)$, as a function of M, is a *quadratic character* which differs from the residue character $M \mapsto \left(\frac{M}{D}\right)$ modulo D by a factor depending only on the degree and on the highest coefficient of M (if D is considered to be fixed). From this he concludes:

   Let $n > 0$ denote the degree of the discriminant D. If M ranges over all monic polynomials, relatively prime to D and of fixed degree m, then

$$\sum_{\deg M = m} \chi_D(M) = 0 \qquad if \qquad m \geq n. \tag{3.3}$$

The condition that M is relatively prime to D can be omitted by putting $\chi_D(M) = 0$ if M has a common divisor with D.

   This becomes a key result in the "analytic" part of Artin's thesis.

3.1.2   The Analytic Part

In the second part of his thesis Artin introduces the zeta function

$$\zeta_R(s) = \prod_{\mathfrak{p}} \frac{1}{1 - |\mathfrak{p}|^{-s}} = \sum_{\mathfrak{a} \neq 0} |\mathfrak{a}|^{-s} \tag{3.4}$$

where s is a complex variable. Here, $\mathfrak{p}$ ranges over all prime ideals $\neq 0$ of $R = K[x, \sqrt{D}]$ and $|\mathfrak{p}| = q^{\deg \mathfrak{p}}$ is the number of elements in the residue field $R/\mathfrak{p}$.[4] Accordingly $\mathfrak{a}$ ranges over all integer ideals $\neq 0$ of R and $|\mathfrak{a}|$ is the order of $R/\mathfrak{a}$. Since R is a Dedekind ring, every nonzero prime ideal $\mathfrak{p}$ of R defines a unique prime divisor of the field F which may also be denoted by $\mathfrak{p}$, and $R/\mathfrak{p} = F\mathfrak{p}$. But in this way one does not obtain all the prime divisors of F. The primes at infinity (where $x(\mathfrak{p}) = \infty$) do not correspond to prime ideals of R. There are one or two such primes of F, according to whether $\sqrt{D}$ is imaginary or real in the sense of Artin. Hence Artin's zeta function $\zeta_R(s)$ does not coincide with the zeta function $\zeta_F(s)$ of the field in (2.2); the latter was defined later by F.K. Schmidt (see Sect. 4.1). $\zeta_R(s)$ and $\zeta_F(s)$ differ by the Euler factors belonging to the one or two infinite places. More precisely: We have

$$\zeta_F(s) = \zeta_R(s) \cdot \prod_{x(\mathfrak{p})=\infty} \frac{1}{1 - |\mathfrak{p}|^{-s}}. \tag{3.5}$$

Thus Artin's zeta function appears in a sense as an "affine" object, belonging to the affine curve with R as its coordinate ring—whereas F.K. Schmidt's function belongs to the nonsingular projective completion of this curve.

Because of this, Artin's formulas look somewhat different from the corresponding formulas which we are used to when referring to F.K. Schmidt's zeta function. Also, the Riemann hypothesis for Artin's zeta function concerns the "nontrivial" roots of $\zeta_R(s)$ only: these are supposed to have real part $\Re(s) = \frac{1}{2}$, whereas there may be some "trivial" zeros of $\zeta_R(s)$. It will be seen below which roots are "trivial" in this sense.

Artin himself in his thesis is not concerned with the difference between $\zeta_R(s)$ and $\zeta_F(s)$; at that time the latter was not yet defined and Artin's zeta function $\zeta_R(s)$ was for him the first and only zeta object to study in quadratic function fields. But in a letter to Herglotz written on 30 November 1921, Artin already considered the change of his zeta function by birational transformations. See page 30 below.

The investigation of the properties of $\zeta_R(s)$ is the main objective in the second part of Artin's thesis. The first and essential observation is that $\zeta_R(s)$ is a rational function if considered as a function of the variable $t = q^{-s}$. If this would have been known for the F.K. Schmidt zeta function ζ_F already then it would follow immediately for Artin's zeta function ζ_R too, because the (one or two) absent Euler factors are also rational functions of t. F.K. Schmidt in [Sch31a] proves the rationality of his zeta function by means of the Theorem of Riemann-Roch. But as we have just said, F.K. Schmidt's result was later, and the Riemann-Roch Theorem

---

[4]For the moment, while discussing Artin's thesis I am denoting the prime ideals of R by gothic letters $\mathfrak{p}$ since the latin capital P has been used already for prime polynomials in $K[x]$. Later I will switch again to the notation introduced in Sect. 2.1 where latin letters like P, Q etc. denote primes of F.

had not yet been established at the time of Artin's thesis for function fields with
finite base field. Thus Artin had to use another strategy. He represented the zeta
function of the quadratic function field by means of an L-series—following closely
the procedure in the number field case.

Let $P \in K[x]$ denote a monic prime polynomial. The decomposition type of P
into prime ideals in R is governed by the value of the character $\chi_D(P)$, as Artin
had shown in the first part of his thesis, see (3.2). According to whether $\chi_D(P) = 1$
or $= -1$, the prime P splits in R into two different prime ideals each of relative
degree 1, or P remains prime of relative degree 2. If $\chi_D(P) = 0$ then P is ramified
in R.

This leads to the following product decomposition where P ranges over all monic
prime polynomials of $K[x]$ and M over all monic polynomials, and where we have
put $t = q^{-s}$:

$$\zeta_R(s) = \prod_P \left(\frac{1}{1 - t^{\deg P}} \right) \left(\frac{1}{1 - \chi_D(P)t^{\deg P}} \right)$$

$$= \sum_{m \geq 0} q^m t^m \cdot \sum_{m \geq 0} \sigma_m t^m \quad \text{with} \quad \sigma_m = \sum_{\deg M = m} \chi_D(M).$$

The second factor is the L-series with respect to the character χ_D. It appears as a
power series in $t = q^{-s}$. (Artin does not introduce the variable t and keeps writing
q^{-s} instead; in this way the power series in t appears as a Dirichlet series in s. He
also does not use the notation L at this point.) But now, using the relation (3.3) of
the arithmetic part, Artin concludes that this power series is in fact a polynomial of
degree $< n = \deg D$. This gives:

Artin's zeta function can be written in the following form:

$$\zeta_R(s) = \frac{1}{1 - qt} \cdot \sum_{m=0}^{n-1} \sigma_m t^m = \frac{L_R(t)}{1 - qt} \quad \text{(with } t = q^{-s}). \tag{3.6}$$

*where $L_R(t)$ is a polynomial of degree $\leq n - 1$ whose coefficients σ_m are given as
above.* (Actually, it turns out with the help of the functional equation that the degree
of the polynomial $L_R(t)$ is precisely $n - 1$).

This result is central in Artin's thesis. It shows explicitly that Artin's zeta function
of a quadratic function field behaves quite differently from the zeta function in the
quadratic number field case. In the introduction to his thesis Artin had said:

"*It appears that a general proof* [of the Riemann hypothesis for quadratic
function fields] *will have to deal with problems of similar type as with
Riemann's $\zeta(s)$, although here* [in the function field case] *the situation is
clearer and more lucid because it essentially concerns polynomials.*"

From the second part of this statement we see that he was fully aware of the fact
that in the function field case the situation is different from the classical case.

Artin was not the first to have observed this. There is a paper by H. Kornblum [Kor19] where L-series $L(t|\chi)$ are considered, for arbitrary (not only quadratic) characters χ in $K[x]$ modulo D.[5] For non-trivial characters Kornblum had shown that $L(t|\chi)$ is a polynomial, and he had done it in the same way as Artin does in his thesis by means of (3.3).

Artin immediately starts to draw consequences from (3.6). Among other things he arrives at the following results:

1. **Trivial and non-trivial zeros**: If $\sqrt{D}$ is "real" then $L_R(1) = 0$. If $\sqrt{D}$ is "imaginary" and the degree of D is even then $L_R(-1) = 0$. These zeros are called "trivial". All the other "nontrivial" zeros of $L_R(t)$ are contained in the region $\frac{1}{q} < |t| < 1$ which means $0 < \Re(s) < 1$.

   The Riemann hypothesis claims that the *non-trivial roots* have real part $\Re(s) = \frac{1}{2}$. Artin cannot prove this in general but he observed that the weaker result in statement **1.** has already important consequences.

2. **Class number formula**: Formulas for the class number h_R of R are obtained by computing the residue of $\zeta_R(s)$ at $s = 1$. Artin obtains the following formulas:

$$\lim_{s \to 1}(s-1)\zeta_R(s) = \begin{cases} \dfrac{\sqrt{q} \cdot h_R}{\sqrt{|D|} \cdot \log q} & \text{if } \sqrt{D} \text{ is "imaginary" and } \deg D \text{ odd} \\[3mm] \dfrac{(q+1)h_R}{2\sqrt{|D|} \cdot \log q} & \text{if } \sqrt{D} \text{ is "imaginary" and } \deg D \text{ even} \\[3mm] \dfrac{(q-1)\rho(R)h_R}{\sqrt{|D|} \cdot \log q} & \text{if } \sqrt{D} \text{ is "real".} \end{cases}$$

(3.7)

In the "real" case, $\rho(R)$ denotes the regulator of R which is the integer defined by the formula $|\varepsilon| = q^{\rho(R)}$ involving the fundamental unit ε. Thus $\rho(R) = \frac{\log|\varepsilon|}{\log q}$. By means of the functional equation (see below) this can be transferred to $s = 0$, i.e., $t = 1$. Artin arrives at the following formulas involving the coefficients σ_m of the polynomial $L(t)$:

$$h_R = \begin{cases} \sigma_0 + \sigma_1 + \cdots + \sigma_{n-1} = L_R(1) & \text{if } \sqrt{D} \text{ is "imaginary"} \\[3mm] -\dfrac{1}{\rho(R)}(\sigma_1 + 2\sigma_2 + \cdots + (n-1)\sigma_{n-1}) = -\dfrac{1}{\rho(R)}L'_R(1) & \text{if } \sqrt{D} \text{ is "real".} \end{cases}$$

(3.8)

---

[5]I have mentioned Kornblum's paper already, see page 15. The young Kornblum had been a Ph.D. student of Landau in Göttingen. He died in early World War I. The manuscript of his Ph.D. thesis had been almost completed; it was edited and commented by his academic teacher Landau and published 1919 in the newly founded *Mathematische Zeitschrift*. Artin cites Kornblum, and he points out that his (Artin's) results on the number of prime polynomials in an arithmetic progression are essentially stronger than Kornblum's.

3. **Functional equation**: I will not write down the explicit form of the functional
   equation for Artin's zeta function. Let us be content by saying that $\zeta_R(1-s)$ is
   expressed in terms of $\zeta_R(s)$, and hence $L_R(\frac{1}{qt})$ in terms of $L_R(t)$. This yields
   certain symmetry relations of the coefficients σ_m which Artin uses to reduce
   computations of the class number h_R by means of (3.8).
4. **Table of class numbers**: Artin computes numerically tables of class numbers,
   for small prime numbers $q = p \leq 7$ and small degrees 3 or 4. (These fields
   are all elliptic). Using the above mentioned symmetry properties it suffices to
   compute σ_1. In each of Artin's cases, about 40 in number, he verifies numerically
   the Riemann hypothesis, by computing the coefficient σ_1 and discussing the
   nontrivial zeros of the corresponding L-polynomial. Furthermore, Artin remarks
   (without showing tables) that for $q = p = 3$ he has extended his computations
   to discriminants of degree 5, except some prime discriminants, and again, the
   Riemann hypothesis could be verified. (These fields are hyperelliptic of genus 2).

   This last remark about discriminants of degree 5 had been inserted into the
   manuscript shortly before the paper was sent to print, which was 13 October
   1921. Hence between the first version, which was submitted to the Leipzig
   Faculty in May 1921, and the final version in October, Artin had done quite a
   number of additional computations in order to verify the Riemann hypothesis.
   But now, in a letter to Herglotz on the same day, 13 October 1921, he writes:

   > "*I believe that more computations are of no use, for they would not lead to
   > a decision. (If so, then in the negative sense.)*"

   From this statement we can see Artin's own attitude towards the question of
   whether the Riemann hypothesis is valid or not. Although formally he leaves
   the question open, admitting that the final decision may turn out to be negative,
   he is putting that possibility into parentheses. And he decided to stop his
   computations. So we may infer that indeed he expected the Riemann hypothesis
   to hold generally.

   Sometimes in the literature it is said that the Riemann hypothesis for hyperelliptic function fields "was first conjectured by Artin in his thesis". But in Artin's
   thesis we do not find any statement which looks like such conjecture. It is true
   that Artin was the first to state the problem and to open the scene. But as far as we
   know, he never came out with a "conjecture" that the Riemann hypothesis would
   be true. The only indication about his personal attitude is contained in the above
   passage in his letter to Herglotz. In this sense, perhaps, Artin can be said to have
   "conjectured" the Riemann hypothesis. But as far as his published thesis is concerned it would be more precise to say that he had "verified" it in some examples.

   Although Artin cannot prove the Riemann hypothesis in general he proves
   the following conditional result:
5. **Finitely many imaginary fields with given class number**: *If the validity of the
   RHp is assumed for quadratic function fields, the following holds: For all $q > 2$
   and $n = \deg D > 2$ there are only finitely many "imaginary" quadratic function
   fields $F = \mathbb{F}_q(x, \sqrt{D})$ with given class number h. Fields with $h = 1$ are possible
   only for $q = 3$. (Probably only one field.)*

Artin mentions this result for $h = 1$ because in the number field case, the determination of all imaginary quadratic fields with class number 1 was a classical problem which at the time of Artin's thesis was not solved. Nine imaginary quadratic number fields of class number 1 were known but it was not yet known whether that list was complete (which it is). Thus Artin wished to put into evidence that in the case of function fields the situation is easier than for number fields.

The "one field" with class number $h = 1$ for $q = 3$ which Artin mentions, is listed in his class number tables; it belongs to the discriminant $D = x^3 - x - 1$. Artin does not say how he arrived at his prediction that this is the only imaginary quadratic function field with $h = 1$ and $q = 3$, neither in the published thesis nor in his letter to Herglotz. Has he just used computational evidence? Artin's prediction was later verified by MacRae [Mac71]. In addition to the Riemann hypothesis MacRae used the Riemann-Roch Theorem. This, as we know from F.K. Schmidt, is equivalent to the functional equation of the zeta function, and the latter was available to Artin for quadratic function fields. Thus in principle, Artin could have done himself the (elementary) computation which MacRae did. But it is not straightforward.

REMARK Recall that Artin works in characteristic > 2. At the time of MacRae's paper [Mac71] it was possible to handle also the case of characteristic 2. MacRae found that in characteristic 2 there are precisely three "imaginary" quadratic fields with $h = 1$. See also Madan and Queen [MQ72].

3.2 Artin's Letters to Herglotz

As we have seen in the foregoing section, Artin's thesis was devoted to the transfer of the arithmetic of quadratic number fields to the function field case with finite base field. This can be regarded as good work for a young Ph.D. candidate but it did not carry any new and unexpected insight. A. Weil reports in his review of Artin's Collected Papers [Wei79b] that at the time when Artin wrote his thesis he was very young and, "*as he used to say later, very ignorant*". In particular Artin did not quote Dedekind-Weber [DW82] which, Weil says, "*would have been more relevant to his purposes.*" That was finally done by F.K. Schmidt, see Sect. 4.1.

The RHp did not play any significant role in Artin's thesis. But the situation changed in the letters of Artin to Herglotz. There Artin reported on some further work in the direction of the RHp, and this pointed the way forward.

The legacy of Gustav Herglotz contains several letters from Artin. Two of those letters, written in November 1921, carry important ideas concerning the Riemann hypothesis. These letters became known through Ullrich's paper [Ull00]. There we can observe in a nutshell several ideas which later became important for the proof of the Riemann hypothesis.

The letters were sent from Göttingen where Artin had moved to as a post-doc. He reports to his former academic teacher about his impressions in the new environment. He also reports about the seminars which he is attending. But otherwise he feels quite lonely:

"Unfortunately, I have here very little contact to the lecturers, and therefore I am missing the personal stimulus which in Leipzig I have had to such high degree by yourself, Herr Professor. For this I will always be grateful to you."[6]

But then he proceeds to report about his work, following the lines set in his thesis.

3.2.1   Extension of the Base Field

Artin finds it self-evident that the whole theory in his thesis, although presented over $\mathbb{F}_p$ as the base field, remains valid over every finite base field K. This I have mentioned already (see page 14). Now, in a letter dated 13 November 1921, he studies base field extensions in a systematic way.

Let $F = K(x, \sqrt{D})$ be a quadratic function field with a finite base field K, and deg $D > 0$. Consider the extension K_r of K of degree r and let $F_r = K_r(x, \sqrt{D})$ be the corresponding base field extension of F. Let R, R_r be the integral closures of $K[x]$ in F, and of $K_r[x]$ in F_r respectively. The corresponding zeta functions are denoted by $\zeta_R(s)$ and $\zeta_{R_r}(s)$. Artin had found a connection between $\zeta_R(s)$ and $\zeta_{R_r}(s)$. His formula is:

$$\zeta_{R_r}(s) = \prod_{0 \le \mu \le r-1} \zeta_R\left(s + \frac{2\pi i \mu}{r \log q}\right) \tag{3.9}$$

Artin concludes:

" 1.) If the Riemann hypothesis holds for $\zeta_R(s)$ then also for $\zeta_{R_r}(s)$. Hence with every zeta function there is a whole bunch of infinitely many other zeta functions for which the Riemann hypothesis holds. (In every F_r with arbitrary r).
2.) Conversely: If the Riemann hypothesis is proved for $\zeta_{R_r}(s)$ then also for $\zeta_R(s)$. Hence it is sufficient to prove the Riemann hypothesis over all those finite fields K for which the discriminant D split into linear factors. Over every such field there are only finitely many cases to be dealt with."

In the next letter, dated 30 November 1921, he goes one step further and obtains the following conclusion. Recall the notation in (3.6) where σ_1 denotes the first

---

[6]Constance Reid reports in her book [Rei76] that Herglotz had little contact with his students. If this was the case then it seems that his relation to Artin was an exception.

coefficient of the polynomial $L_R(t) = 1 + \sigma_1 t + \cdots$. Artin derives from (3.9) the following criterion for the validity of the Riemann hypothesis:

"...*for the proof of the Riemann hypothesis it is only necessary to have the following "raw" estimate of σ_1, but over* all *Galois fields:*

$$|\sigma_1| < A\sqrt{q}$$

where A depends only on the degree of the discriminant."

It is seen from the definition in (3.6) that

$$\sigma_1 = N_R - q \qquad (3.10)$$

where N_R is the number of prime ideals of degree 1 in the ring R. Thus we may write the above inequality as $|N_R - q| < A\sqrt{q}$. But this should be known not only for q but also for all powers of q, and with the same bound A. So we should better write Artin's criterion as

$$|N_{R_r} - q^r| < A\sqrt{q^r} \qquad \text{for } r = 1, 2, 3, \ldots \qquad (3.11)$$

where the bound A does not depend on r.

We see that now Artin begins to think about a general proof of the Riemann hypothesis, more than just numerical verification in examples. In fact, the criterion (3.11) has become crucial in all later setups for the proof of the Riemann hypothesis. It is remarkable that Artin at this early stage is fully aware of the importance of this criterion. I shall give Artin's proof later in the context of F.K. Schmidt's birationally invariant zeta function (see page 52).

3.2.2   Complex Multiplication

Suppose the discriminant $D = D(x)$ is cubic. Then all the zeros of the quadratic polynomial $L_R(t) = 1 + \sigma_1 t + qt^2$ are non-trivial in the sense as explained above. Consider the "inverse" polynomial

$$L_R^*(t) = t^2 + \sigma_1 t + q. \qquad (3.12)$$

According to what Artin has just found, the RHp is equivalent to the inequality

$$\sigma_1^2 - 4q < 0 \qquad (3.13)$$

which is to say that the roots of $L_R^*(t)$ are imaginary.

In this situation Artin puts himself the question how to describe the connection between the elliptic function field $F = K(x, \sqrt{D})$ over a finite field K of order q, and the square free kernel $d \in \mathbb{Z}$ of $\sigma_1^2 - 4q$, so that $\sigma_1^2 - 4q = u^2 d$ with $u \in \mathbb{Z}$. It appears that he hoped to find some way to deduce that $d < 0$ and, hence, to prove the RHp at least in the case of a cubic discriminant D. (Artin writes $\sigma_1^2 - 4q = -u^2 d$ and so he is looking for a proof of $d > 0$).

Later, when Hasse was going to prove the Riemann hypothesis for elliptic function fields F (i.e., for cubic discriminants D), one of the essential steps in his proof was the discovery that to each elliptic field F there belongs a so-called endomorphism ring such that every member of this ring satisfies a quadratic polynomial with negative discriminant. The polynomial $L_R^*(t)$ above is the quadratic polynomial which belongs to the so-called "Frobenius operator" of $F|K$. This is the theory of "complex multiplication" in characteristic p. (See Sect. 7.4).

It seems remarkable that Artin in 1921, without referring to complex multiplication and Frobenius operator, poses the problem in this way. Now we can understand that in 1934, when he invited Hasse for a lecture course in Hamburg, he was so enthusiastic about Hasse's work. (See page 90.) For Hasse could do what Artin had seen in 1921 but was not able to prove in general. Probably Hasse did not know about Artin's 1921 letter to Herglotz. In that letter, Artin had written:

"The big question is, which d belongs to the given discriminant D? In general I know nothing about this. But in two cases I can decide this and will now report on it."

And he proceeds to discuss the following two cases:

$$D = x^3 - x \quad \text{when} \quad q \equiv 1 \mod 4$$

$$D = x^3 - 1 \quad \text{when} \quad q \equiv 1 \mod 6.$$

For the proof of the RHp the respective congruence condition for q is not essential because of what Artin has just mentioned: One may replace the base field K by its extension of order 2, hence q by q^2, which then satisfies the congruence condition (if characteristic $p > 3$). Also, one may replace $x^3 - x$ by $x^3 - bx$ with $b \in K$, because after a suitable extension of K this can be transformed easily into the form as given. Similarly, $x^3 - 1$ may be replaced by $x^3 - b$.

In those two cases Artin succeeds to determine $d = -1$ and $d = -3$ respectively. And so he can proudly announce:

"Accordingly one has the Riemann hypothesis for all q, and all $D = x^3 - bx$ and $D = x^3 - b$."

Note that this result is of quite different nature from the numerical verification which Artin gave in his published thesis, because it is valid for *all* q whereas in the thesis only small prime numbers are involved.

We observe that the two cases which Artin had chosen are elliptic fields of absolute invariant $j = 0$ and $j = 1$, and these are known to have endomorphism ring

$\mathbb{Z}[i]$ and $\mathbb{Z}[\varrho]$ respectively, where i is a primitive 4-th and ϱ a primitive 3-rd root of unity. In these cases the endomorphism ring has units of order > 2. But it seems that Artin at that time was not aware of the connection to complex multiplication since he did not mention it in his letter. But then, how did he proceed to obtain $d = -1$ and $d = -3$ in those two cases?

He did it with what he called a trick. Consider $D = x^3 - x$ with the condition $q \equiv 1 \mod 4$. Artin compares the field $F = K(x, \sqrt{D})$ with the field $\overline{F} = K(x, \sqrt{\overline{D}})$ where $\overline{D} = x^3 - bx$, and b is not a square in K. These fields are not isomorphic but they become isomorphic after base field extension from K to K_4, the extension of degree 4 over K. Let $\overline{\sigma}_1$ be the first coefficient of the L-polynomial for $\overline{F}$, in the same way as σ_1 is defined for F. Then Artin compares F_2 and $\overline{F}_2$, the base field extensions of degree 2. Using (3.9) he finds the following relation:

$$\sigma_1^2 + \overline{\sigma}_1^2 = 4q \qquad (3.14)$$

which shows that, indeed, $\sigma_1^2 - 4q = -\overline{\sigma}_1^2$ and thus $d = -1$.

(A closer look at this "trick", which today could be interpreted in the framework of Galois cohomology, will reveal that it depends on the existence of automorphisms of F of order > 2. Thus the two examples which Artin used are essentially the only ones for which this works).

By the very definition of σ_1 and $\overline{\sigma}_1$ Artin obtains the following cute side result in the case when $q = p$ is a prime number: Let $p \equiv 1 \mod 4$. Then: *If b is a non-square modulo p and if we put*

$$\alpha = \sum_{0 \leq \nu \leq \frac{p-1}{2}} \left(\frac{\nu^3 - \nu}{p} \right) \quad \text{and} \quad \beta = \sum_{0 \leq \nu \leq \frac{p-1}{2}} \left(\frac{\nu^3 - b\nu}{p} \right) \qquad (3.15a)$$

then

$$p = \alpha^2 + \beta^2 . \qquad (3.15b)$$

Here, the brackets denote the ordinary quadratic residue symbol in $\mathbb{Z}$.

Thus Artin's arguments lead to an explicit algorithm to write any $p \equiv 1 \mod 4$ as a sum of two squares. He writes, however:

"... *unfortunately, this result is not new, it had been found by Jacobsthal using pure computation, as I have been informed some days ago.*"

Artin does not give any reference. A search in the *Jahrbuch* database reveals that the result is implicitly contained in Jacobsthal's Berlin thesis 1906. The publication appeared one year later in Crelle's Journal [Jac07].

But, Artin added, by doing a similar trick for the discriminant $D = x^3 - 1$ he had obtained the following which he believed was new: Let $p \equiv 1 \mod 6$. Then: *If b is not a cubic residue modulo p and if we put*

$$\alpha = \sum_{0 \le \nu \le p-1} \left(\frac{\nu^3 - 1}{p} \right) \qquad \text{and} \qquad \beta = \sum_{0 \le \nu \le p-1} \left(\frac{\nu^3 - b}{p} \right) \qquad (3.16a)$$

then

$$3p = \alpha^2 - \alpha\beta + \beta^2. \qquad (3.16b)$$

From this one can easily obtain a representation of p by the quadratic form $X^2 - XY + Y^2$.

3.2.3 Birational Transformation

Artin considers applying a linear fractional transformation to x. He replaces x by some $x' = \dfrac{ax + b}{x - c}$ with $a, b, c \in K$. Let R' denote the integral closure of $K[x']$ in F. Artin gives a formula expressing the ζ-function belonging to R' in terms of the ζ-function belonging to R. We do not write down this formula because it is quite obvious, exchanging the Euler factors belonging to the infinite primes with respect to x, with the Euler factors belonging to the infinite primes of x'. Artin remarks:

> "*If D is divisible by x − a and of even degree then the degree will decrease. Since D will obtain a linear factor after suitable increase of the base field K, one may assume the degree to be odd. This decrease of degree looks very similar to hyperelliptic integrals.*"

By "degree" he means the degree of the discriminant D. At the same time, the degree of the corresponding L-polynomial, which is one less than the degree of D, will also decrease.

By considering such transformations, Artin gets rid of the limitations which are set by the attempt to obtain complete analogy to the number field case. For, in the number field case the ring $\mathbb{Z}$ is given and cannot be changed; if one insists on complete analogy then in the function field case, the polynomial ring $K[x]$ should be fixed and should not be changed. By leaving this restriction behind and considering also linear fractional transformations, Artin starts to use fully the possibilities which function fields offer, which are not available in number fields.

In this way Artin can simplify the problem by reducing it to the case where the degree of D is odd. Then $\sqrt{D}$ is "imaginary" and remains so after every base field extension; thus for the proof of the Riemann hypothesis it is not necessary to consider the "real" case any more. Although Artin does not mention it, he certainly knew that if $\deg D$ is odd then all the zeros of the polynomial $L_R(t)$ are nontrivial and they are birational invariants of the field F.

This is the first step towards a birational invariant definition of the zeta function, which later was given by F.K. Schmidt.

3.3 Hilbert and the Consequences

By all what Artin wrote in his letters to Herglotz it is evident that he had become quite interested in the Riemann hypothesis at least for quadratic function fields. Surely, there is a long way to go from the knowledge that (3.11) would be sufficient, to an actual proof. We shall see that it required an enormous effort until this goal was reached. But certainly his criterion (3.11) is to be considered as a first important step.

Why did Artin not proceed further in this direction? I have already narrated in Sect. 1.2 that it was Hilbert's reaction which had the effect that he left this topic. This is evidenced by the editors of his "Collected Papers" [Art65] who put his thesis quite singularly into a separate chapter, thus indicating that the thesis is somewhat isolated among the other papers of Artin, none of which can be regarded to be closely related to the thesis.

But Artin continued to be interested in the subject and observed keenly the further development; this can be inferred, e.g., from his letter to Hasse cited on page 90. In Artin's lectures on number theory he included function fields. He was striving for a unified theory for number fields and function fields, which later was established as the theory of "global fields". See, e.g., his famous paper jointly with Whaples [AW45]. But in his publications Artin kept silent on the topic of the Riemann hypothesis in characteristic p.

As Ullrich [Ull00] reports, Herglotz had been able to arrange that Artin could move from Göttingen to Hamburg where he was offered by Blaschke a position as assistant.[7] There, in the very neighborhood of Hecke, Artin found the mathematical atmosphere which suited him. Soon he rose to one of the leading figures in algebra and number theory. But for the RHp his thesis remained his only published contribution.

Summary

Artin's thesis is concerned with quadratic function fields over finite base fields. Artin goes about to transfer the arithmetic theory of quadratic number fields to the case of characteristic p. This includes, firstly the arithmetic properties like decomposition

---

[7] I would like to use this opportunity to point out that it was Blaschke, the first mathematician at the newly founded University of Hamburg, who succeeded to raise this place within a few years to one of the leading mathematical centers in Germany. He did this through a careful *Berufungspolitik*. The Hamburg Mathematical Seminar in its first decades is a good example that mathematical excellence cannot be created by more money or more positions only, but that the decisive point is to attract excellent people.

into prime ideals, theory of units, class number etc., and secondly the analytic theory of the zeta function of a quadratic function field, in analogy to quadratic number fields. Artin observes that his zeta function is a rational function of $t = q^{-s}$, and it is a polynomial in t up to a trivial factor. Artin derives the class number formula in terms of the zeros of this polynomial and he computes it in a number of about 30 numerical cases, most of them elliptic function fields. On the way he observes that in those cases the RHp holds for his zeta function. But otherwise there is no indication that Artin was particularly interested in the RHp when he wrote his thesis.

After having received his degree, Artin spent a year as post-doc in Göttingen. There, in several letters to his academic teacher Herglotz in Leipzig, Artin developed some further results in which we can observe the nucleus of what later will become essential features in the general proof of the RHp. From this we can deduce that meanwhile Artin had become more interested in the RHp. But these further results were never published. The reason was Hilbert's harsh criticism of Artin's work when Artin gave a colloquium talk about it in Göttingen. (I have reported on this in the foregoing chapter already). Nevertheless Artin continued to be keenly interested in the development of the theory of algebraic function fields.

3.4 Side Remark: Gauss' Last Entry

There is one paper of Herglotz which directly concerns the RHp for a function field and therefore has to be mentioned here [Her21]. That paper appeared in the same year when Artin had submitted his thesis, and its title is: "*On the last entry in Gauss' diary.*" This concerns the integer solutions of the congruence

$$x^2 y^2 + x^2 + y^2 \equiv 1 \mod p \tag{3.17}$$

for a prime number $p \equiv 1 \mod 4$. Gauss, in his diary dated 9 July 1814, had noted the number of solutions as follows:

Decompose p in the ring $\mathbb{Z}[i]$ of Gauss integers as $p = \pi \pi'$, as a product of an "imaginary" prime number π and its conjugate π'.[8]

We may identify the factor rings $\mathbb{Z}[i]/\pi = \mathbb{Z}/p$, and hence the solutions of (3.17) in $\mathbb{Z}$ (mod p) are the same as the solutions in $\mathbb{Z}[i]$ (mod π) of

$$x^2 y^2 + x^2 + y^2 \equiv 1 \mod \pi . \tag{3.18}$$

---

[8]There is a clash of notation in the mathematical literature, and also in this book. In Analysis the greek letter "π" is used to denote the real number which is half of the circumference of the unit circle. (See formula (3.9) on page 26). In the present context "π" denotes an algebraic number, namely a factor of p in an imaginary quadratic number field. Later when I will discuss Hasse's theory of complex multiplication in characteristic p the letter "π" will denote the Frobenius operator. (See Sect. 7.4.4). I believe it will always be clear from the context which "π" is meant at the time.

Now, Gauss had found that the number of solutions of (3.18) equals the norm (from $\mathbb{Q}(i)$ to $\mathbb{Q}$):

$$\mathcal{N}(\pi - 1) = (\pi - 1)(\pi' - 1) = p - \mathcal{S}(\pi) + 1, \qquad (3.19)$$

with the following specifications:

1. The 4 infinite solutions $(x, y) \mapsto (\infty, \pm i)$, $(x, y) \mapsto (\pm i, \infty)$ have to be included in the counting.
2. π has to be normalized such that $\pi \equiv 1 \mod (1 - i)^3$.

This normalization can be achieved by multiplying π with a suitable unit $\varepsilon \in \{\pm 1, \pm i\}$. After such normalization, π is called "primary". Note that $1 - i$ is a Gauss prime number dividing 2.

Gauss did not actually give a proof, he had observed this fact "by induction" which according to the terminology at his time meant either heuristically, or experimentally for several p.

The story of this "last entry" is well known; see, e.g., the presentation in Lemmermeyer's book [Lem00] and the literature cited there. Herglotz presents a proof of Gauss' statement. Before discussing this, let us first point out the connection of Gauss' statement to the RHp for the function field $F = K(x, y)$ where x, y satisfy the lemniscate equation

$$x^2 y^2 + x^2 + y^2 = 1. \qquad (3.20)$$

Thereby $K = \mathbb{F}_p$ is assumed to be the prime field.

F is a quadratic extension of the rational function field $K(x)$, and from (3.20) we obtain

$$y^2 = \frac{1 - x^2}{1 + x^2}. \qquad (3.21)$$

This relation is satisfied by the 4 infinite solutions mentioned above. This explains why these are also counted as solutions of (3.18). Seen in this way, Gauss' statement says that $\mathcal{N}(\pi - 1)$ is *the number of all places of degree 1 of F*; this number is independent of the choice of generators x, y of F.

If we multiply the above relation with $(1 + x^2)^2$ we find

$$F = K(x, \sqrt{1 - x^4}) \qquad (3.22)$$

which is of the form Artin had discussed in his thesis, with $D = 1 - x^4$. We have seen in Sect. 3.2.3 that Artin admits also linear fractional transformations in order to reduce the degree of the discriminant D. Such a degree reduction is obtained by putting $x = \frac{1 + ix'}{1 - ix'}$ where i is a primitive 4-th root of unity in K. We compute

$$F = K(x', \sqrt{x'^3 - x'}) \qquad (3.23)$$

This is precisely the form which Artin had discussed in his letter to Herglotz on 30 November 1921; see Sect. 3.2.2. In that letter Artin had presented his proof of the Riemann hypothesis for this field F.

But the validity of the Riemann hypothesis for this field is also immediate from Gauss' statement, which is seen as follows:

As Artin had pointed out in his letter to Herglotz, the Riemann hypothesis is equivalent with the inequality (3.13) which, according to (3.10), can be written as

$$|N_{R'} - p| < 2\sqrt{p}. \tag{3.24}$$

Here, $N_{R'}$ denotes the number of prime ideals of degree 1 in the integral closure R' of $K[x']$ in F. Every such prime ideal defines a place of F of degree 1. There is precisely one more place of F of degree 1, and this belongs to $x' \mapsto \infty$. (This is so since the degree of the discriminant $D' = x'^3 - x'$ is odd). Thus the number of all places of F of degree 1 is $N_{R'} + 1$. Hence, using Gauss' statement, we obtain $N_{R'} + 1 = \mathcal{N}(\pi - 1) = p - \mathcal{S}(\pi) + 1$. We conclude that $N_{R'} - p = -\mathcal{S}(\pi)$. Writing $\pi = a + bi$ we have $\mathcal{S}(\pi) = 2a$ and since $p = a^2 + b^2$ we see that $\mathcal{S}(\pi)^2 = 4a^2 < 4p$ which gives (3.24).

Conversely, from Artin's results as explained in his letter to Herglotz, it is straightforward to deduce Gauss' statement (3.19) for a Gauss prime π dividing p, without however the normalization in 2. which is a more subtle affair.

In any case, we see that between the results of Herglotz and those of his disciple Artin there is a very close connection indeed. The question arises: Why does none of them cite the work of the other?

As pointed out above, Herglotz's paper [Her21] appeared in the same year as Artin submitted his thesis. It is conceivable that Herglotz had completed his work before Artin started his thesis. And that Herglotz, because he had seen the connection with the RHp for one particular quadratic function field, had proposed to Artin to look into this question quite generally, for arbitrary quadratic function fields. But then Artin should have cited Herglotz's paper as the one where his work started. Since he did not, and since he explained the case in such detail in his letter to Herglotz, we conclude that Artin did not know Herglotz's paper at that time. Why not? Why did Herglotz not show his paper to Artin? The theory of complex multiplication which Herglotz used, was to become in the future, in the hands of Hasse, a powerful tool for the proof of the RHp for arbitrary elliptic function fields. If Herglotz would have foreseen the importance of complex multiplication for the Riemann hypothesis then he would probably have proposed to Artin to study that theory, with the aim to apply this to the proof of RHp. But apparently he did not.

Another possibility would be that Herglotz wrote his paper later, after he had seen Artin's thesis. Inspired by Artin's work he may have realized that Gauss' last entry would imply the validity of the RHp for the function field of the lemniscate, and so he set out to work on it. In this case Herglotz would have cited Artin and shown the connection between his paper and the work of his disciple Artin. But he did not.

I have no obvious explanation for the fact that none of them, Herglotz and Artin, cited the other.

As to the contents of Herglotz's paper [Her21]: He first presents the statement (3.19) of Gauss' last diary entry, and then reports (without giving any reference) that Dedekind had verified Gauss' statement for all $p < 100$, and that Fricke had pointed out the coincidence of the Eq. (3.20) with the equation which is satisfied by the lemniscate functions of Gauss[9]:

$$x = \varphi(u) := \sin\operatorname{lemn}(u) \qquad \text{and} \qquad y = \psi(u) := \cos\operatorname{lemn}(u).$$

Herglotz continues:

"In addition we shall remark here that the solutions of (3.17) coincide precisely with the congruence solutions modulo π of the division equations for

$$x = \varphi\left(\alpha\frac{\omega_3}{\pi-1}\right) \qquad \text{and} \qquad y = \psi\left(\alpha\frac{\omega_3}{\pi-1}\right)." \tag{3.25}$$

(Herglotz uses the Weierstrass notation, namely: $\omega_3 = -(1+i)\omega$ where ω is the primitive real period of $\varphi(u)$. In the above equation, $\alpha \in Z[i]$ is arbitrary).

After this introduction Herglotz starts with his calculations. The division equation is used in a form which Weierstrass had given, and the greater part in Herglotz's calculations seems to be catching the connection to the notations which Weierstrass had used. We have not checked these calculations. Schappacher [Sch97] writes that *"Herglotz uses the Weierstrass theory, albeit with notation that has not quite survived to the present day."* The essential feature in these calculations is the observation that at a certain point there appears an Eisenstein equation with respect to the prime π and therefore, modulo π, the result follows.

The normalization $\pi \equiv 1 \bmod (1-i)^3$ implies that the multiplication with π of the arguments in (3.25) induces the map $(x, y) \to (x^p, y^p) \bmod \pi$, which is to say the Frobenius map of the reduced curve. This is implicitly used in Herglotz's computations.

We observe that Herglotz never says explicitly that Gauss' statement follows from his results. He seems to consider it as self-evident that the number of distinct (x, y) in (3.25) is $\mathcal{N}(\pi - 1)$ when α ranges over $\mathbb{Z}[i]/(\pi - 1)$. Indeed this belongs to the basics of complex multiplication and we have no doubt that Herglotz regarded it as such. He was a very knowledgable mathematician, well acquainted with the old masters of the science. When he not explicitly claimed to have proved the statement in Gauss' last entry, then because in his eyes this was self-evident from what he

---

[9]The function $\sin\operatorname{lemn}$ is defined as the inverse function of $\int_0^x \frac{dx}{\sqrt{1-x^4}}$, and accordingly $\cos\operatorname{lemn}$ by means of (3.21).

really proved. His style used to be very concise—or maybe we should say "minimal" if it comes to explanations for the reader.

This may have been the reason why the paper of Herglotz was not properly appreciated by the mathematical public of the time. The paper was reviewed by Gábor Szegö in the "*Fortschritte der Mathematik*", and he just repeats the author's claim that the solutions of (3.17) coincide precisely with the congruence solutions modulo π of the division equations for the lemniscate functions. No mention of the fact that this implies the validity of Gauss' statement. The same we can observe in the commentaries in Herglotz's "Collected works" [Her79]: The article on Gauss' last entry was commented on by Theodor Schneider, and again he does not mention that the validity of Gauss' entry follows from Herglotz's result.

When in the year 1933, Hasse used complex multiplication to prove the RHp for all elliptic function fields, he did not cite Herglotz's paper although Herglotz had used the same method, namely complex multiplication, in the special case of the lemniscate function field. And in the joint paper of Hasse and Davenport [DH34] we find an extra section where a particularly simple proof of the Riemann hypothesis for the field $K(x, \sqrt{1 - x^4})$ is given; in view of (3.22) this is precisely the field of the lemniscate. Again, Herglotz's paper was not cited. Probably Hasse (and Davenport too) did not know it.

So we see that Herglotzs' paper was not widely known among the mathematicians of the time, at least it was not realized that it yields a proof of Gauss' last entry. Only after many years this consequence of Herglotz's paper was re-discovered by Deuring [Deu41a]. That paper was dedicated to Herglotz on his 60th birthday in 1941. Deuring mentions Hasse's work on the Riemann hypothesis for elliptic fields and then writes:

> "*Apparently the mathematicians who work on these problems did not notice that this theorem about congruences of genus 1 had been known to Gauss already, at least in the case of lemniscate functions... Finally Herglotz has proved Gauss' statement. His method, namely to use the division of elliptic functions by $\pi - 1$, is the same, in principle, which Hasse had used in his first paper on the Riemann hypothesis.*"

Of course, Deuring knew the theory of complex multiplication well; after all he was the one who, following Hasse's lines, completely remodelled the theory replacing the analytic framework with a purely algebraic one. (See Sect. 8.5). Thus for him it was obvious that Herglotz's results imply the validity of Gauss' last entry. In this way Herglotz's paper came to be appreciated belatedly as a forerunner of Hasse's general theory of elliptic function fields and the RHp.

But even so, I have no explanation why Herglotz himself did not mention his own paper, not to Hasse and not to anyone around Hasse.

On 10 January 1933 Hasse delivered a colloquium talk in Göttingen at the *Mathematische Gesellschaft*. At that time Herglotz was in Göttingen already, having accepted in 1925 an offer from Göttingen as the successor of the applied mathematician C. Runge. It seems to us very probable that Herglotz, being a member of the *Mathematische Gesellschaft*, attended Hasse's talk where Hasse presented his new view of the Riemann hypothesis in connection with the work of Davenport and Mordell. Would it not be natural that in the ensuing discussion, Herglotz would mention that he had proved the RHp for the lemniscate, in his paper on Gauss' last entry? Later in the same year, on 10 December 1933, Hasse gave another talk in Göttingen; at this time he had already obtained his proof of the RHp for arbitrary elliptic curves. This time we know for certain that Herglotz attended the talk for he, together with F.K. Schmidt, had invited Hasse to Göttingen. Herglotz had offered that Hasse may stay in his house during the time of his visit. (I do not know whether Hasse accepted this offer). And again Herglotz did not mention his own paper. If he would have then Hasse, who generally was very careful in his citations, would certainly have mentioned it in one of his papers.

In the summer of 1934 Hasse accepted a position in Göttingen and from then on the two, Hasse and Herglotz, worked as colleagues at the same institution. It is inconceivable that Herglotz was not informed about Hasse's work at that time. Then, why did he keep silent about his own work, realizing that Hasse did not know it?

I have no explanation, except perhaps that the years of Herglotz's activities in number theory had passed and his interest had shifted to partial differential equations and differential geometry. And he had forgotten his own paper on Gauss' last entry. This seems not to be entirely implausible. There are several mathematicians who were known to forget their own work after some time. For instance, Hilbert is said to have completely forgotten his earlier work as soon as he turned to a new field of mathematical research.

Summary

In 1921, the same year when Artin submitted his thesis, Herglotz published a note containing a proof of the statement in the last entry of Gauss' diary of 1814. This concerned the number of solutions modulo p of the lemniscate equation, for $p \equiv 1$ mod 4. It implies the validity of the RHp for the function field of the lemniscate. But Artin does not cite Herglotz's paper in his thesis, and it appears that he did not know it.

The method which Herglotz used belongs to the theory of complex multiplication. Later in 1933, when Hasse obtained the first proof of the RHp for elliptic function fields, Hasse used the same method as Herglotz had used in the special case of the lemniscate function field. But Hasse did not cite Herglotz's paper and it appears he did not know it either. And again, Herglotz did not tell him about it. Quite generally, Herglotz's paper seems not to have been widely known at that time, and it was 1941 only that Deuring brought it to the attention of the mathematical public.

Chapter 4
Building the Foundations

4.1 F.K. Schmidt

© Springer Nature Switzerland AG 2018
P. Roquette, *The Riemann Hypothesis in Characteristic* p
in Historical Perspective, Lecture Notes in Mathematics 2222,
https://doi.org/10.1007/978-3-319-99067-5_4

Friedrich Karl Schmidt (1901–1976) studied at the University of Freiburg where he did his Ph.D. examination in May of 1925. In the introduction to his thesis he refers to Artin's thesis on quadratic function fields which had just appeared in print, and he says:

> *"The analogy to the quadratic number fields which he [Artin] found, suggested the question whether there exists a similar parallelism if one replaces the quadratic by an arbitrary algebraic extension. The present paper shows that indeed this is the case."*

In other words: F.K. Schmidt[1] aimed at generalizing Artin's thesis from quadratic function fields to arbitrary function fields over a finite base field. Formally the thesis advisor of F.K. Schmidt was Alfred Loewy but in reality the topic of F.K. Schmidt's thesis was proposed by Wolfgang Krull (1899–1971) who at that time held a position at the university of Freiburg as *Privatdozent* and assistant to Loewy. Krull had been in Göttingen as a postdoc for two years, where he got into closer contact with Emmy Noether. It appears that it was Emmy Noether who had the idea to generalize Artin's thesis and had challenged Krull to do so, or at least have one of his Ph.D. candidates do it. (Noether reports in a letter to Alexandroff, written in the year 1930, that she had been addressed as "grandmother" in regard to F.K. Schmidt, see [Tob03]).

As said already in Sect. 3.1 Artin's thesis had been divided into two parts, an "arithmetic" and an "analytic" part. In the second part Artin had introduced his zeta function as a complex analytic function and studied its properties. The first part contained those results on quadratic function fields which could be obtained without the analytic zeta function.

Although F.K. Schmidt had announced in the introduction of his thesis that he is going to generalize Artin's thesis for arbitrary function fields, he did not complete this fully. He generalized Artin's first part only. More precisely: Artin's statements (i)–(vi) on quadratic function fields which I have listed on page 18 were generalized to arbitrary function fields $F|K$ over finite base fields, in the following situation: $F|K(x)$ is separable and R the integral closure of $K[x]$ in F. But F.K. Schmidt left no doubt that he will be going to deal with the second part soon. Indeed he did, as we shall see.

REMARK If $F|K(x)$ is separable then x is called a "separating element" for $F|K$. F.K. Schmidt shows that there exist such separating elements provided the base field K is finite or, more generally, K is perfect. Artin in the case of quadratic function fields had avoided this by supposing from the start that the characteristic is not 2.

---

[1]In Germany the name "Schmidt" is quite common. There are several well known mathematicians with this name. In order to identify them it is common to use their first names or first name initials. We shall follow this habit here too; this is the reason why we always use the initials when mentioning F.K. Schmidt, whereas with other mathematicians the initials are not used in general. We shall meet later another Schmid, namely H.L. Schmid (this time without "t").

Later in the same year (1925) the young F.K. Schmidt attended the annual meeting of the German Mathematical Society (DMV) which took place in September in the town of Danzig. There he attended Hasse's famous 1-hour lecture on the new face of class field theory à la Takagi; the lecture notes were published later in an extended form as Hasse's "Class Field Report" [Has26a].

Hasse's lecture, delivered on 15 September 1925, presented the new class field theory for the first time to a broader mathematical audience. This attracted a great deal of attention of the mathematical community.[2] The young F.K. Schmidt too became enthusiastic about the new vista which class field theory opened for algebraic number theory. And he decided on the spot that he would transfer class field theory from number fields to global fields of characteristic $p > 0$. It appears that Hasse had encouraged him to do so. In any case the correspondence between Hasse and F.K. Schmidt, mainly about function fields, started soon after the Danzig meeting and continued for many years.

But it was clear to F.K. Schmidt that first of all Part 2 of Artin's thesis would have to be generalized, in particular the zeta function and the L-series for residue characters of divisors. At that time these objects were considered as indispensable when dealing with class field theory. And he started to work on this.

In a letter of F.K. Schmidt to Hasse dated 6 May 1926—half a year after the Danzig meeting—he writes that he did not yet succeed to get the limit formula for the zeta function in characteristic p. (This formula gives the residue of the zeta function at its pole for $s = 1$ and it was expected that it would contain important information on the class number, similarly as in Artin's thesis in the case of hyperelliptic fields). But later, on 8 August 1926, he informs Hasse that finally he has obtained that formula. (See (4.14) on page 48). In the same letter he announces a prepublication in the "*Erlanger Berichte*". This was a journal published by the University of Erlangen containing research articles of its members from all fields, not only mathematics. It was not widely distributed and would probably not have been read by many mathematicians; its main purpose seems to have been to secure priority for the authors. (This was important for young F.K. Schmidt since in the meantime he had learned about the paper [Sen25] by P. Sengenhorst, a student of Landau, covering the same topic as F.K. Schmidt's thesis. In view of this, F.K. Schmidt did never publish his thesis. But the text is still available as a hand written document at the archives of Freiburg University).

When I read this prepublication [Sch27] I found that F.K. Schmidt originally worked with Artin's zeta function $\zeta_R(s)$ belonging to a finitely generated Dedekind ring R of the function field and its prime ideals. But there is an important appendix

---

[2]It seems that the young group theorist Otto Schreier, who also attended the Danzig meeting, had missed Hasse's talk. For he wrote to his friend Karl Menger one day later, i.e., on 16 September 1925, that scientifically almost nothing of interest had happened during the meeting [Sch14]. Did he also miss Emmy Noether's talk where she for the first time introduced group representations by means of algebras and representation modules [Noe25]? Did he not meet Chebotarev there for whose famous density result he, Schreier, provided a substantial simplification one year later [Sch26] and which inspired Artin's proof of the reciprocity law in class field theory?

"Added in Proof", dated October 1926, in which he switches from Artin's zeta function to the birationally invariant $\zeta_F(s)$ whose definition I have given on page 11. He refers to the well known paper by Dedekind-Weber [DW82] where the birationally invariant viewpoint for the algebraic theory of function fields was proposed for the first time. In that appendix F.K. Schmidt sketches how his formulas, in particular the limit formula, look like when using this birationally invariant zeta function and announces a more detailed exposition.

Thus already in October 1926 F.K. Schmidt gave the birationally invariant definition (2.2) of the zeta function of a function field with finite base field.

I do not know who had suggested this idea to F.K. Schmidt. It may have been Krull who, as mentioned before, kept good relations to Emmy Noether in Göttingen. About Emmy Noether it is known that she admired the work of Dedekind and was promoting his point of view in algebra and number theory.

The announced more detailed version of F.K. Schmidt's paper is contained in his *Habilitation* thesis 1927 and appeared 1931 in the *Mathematische Zeitschrift* [Sch31a] under the title:

"Analytic Number Theory in characteristic p ."

This is the paper where F.K. Schmidt generalizes the second part of Artin's thesis. And he did it in the style of the classical paper of Dedekind-Weber where the theory is built in a birationally invariant way. So he did what A. Weil later said Artin should have done in the first place. (See page 25).

Let me repeat that F.K. Schmidt did not write his paper with the RHp in mind. I have not found any evidence that he was particularly interested in the RHp. His aim was to provide the necessary tools for building class field theory in characteristic p, and it was for this purpose that he developed the properties of the zeta function contained in [Sch31a].

REMARK 4.1 F.K. Schmidt did not fully succeed to build class field theory in characteristic p. In his article [Sch31b] he developed the theory of abelian extensions of global fields in characteristic p under the restrictive assumption that the degree is not divisible by the characteristic. It turned out that this theory can be obtained in complete analogy to the classical case which concerns abelian extensions of number fields, including the theory of L-series $L(s|\chi)$ for ray class characters χ of order prime to p. In his treatment F.K. Schmidt kept close to Hasse's class field report for number fields [Has26a, Has27]. But he did not consider Artin-Schreier extensions nor did he transfer Artin's general reciprocity law to characteristic p. This was later done by Hasse [Has34e] (compare the remark on page 73). It appears that originally F.K. Schmidt regarded his article as a prepublication only which he wished later to adapt to the new developments of class field theory due to Artin, Hasse, Chevalley and others. But he never did so, and today his article on class field theory in characteristic p seems to be forgotten.

REMARK 4.2 F.K. Schmidt's paper [Sch31a] has been refereed in the *Zentralblatt*. There the referee said that for any function field with finite base field

> *"there exists, according to Artin, an ideal theory, a theory of units and an analytic theory, with results which are largely identical with the known properties in algebraic number fields."*

The referee was van der Waerden who had been present in Artin's Hamburg lectures on algebraic number theory. (The famous book "Modern Algebra" is an outgrow of the cooperation of van der Waerden and Artin while working on the print version of Artin's lectures). Thus van der Waerden was well informed when he reported that Artin himself had already achieved the generalization of his thesis to arbitrary function fields with finite base fields (although not published). But van der Waerden admitted that F.K. Schmidt's paper gives a *"new justification"* of the theory, based on divisors and divisor classes instead of ideals and ideal classes. In particular he says explicitly that F.K. Schmidt replaces Artin's zeta function by a new birationally invariant zeta function. I am mentioning this since it confirms what I have said in Sect. 3.3, namely that Artin continued to be interested in function fields with finite base fields, although in his publications he never took up the RHp.

I am now going to report on F.K. Schmidt's paper [Sch31a].

4.2 Zeta Function and Riemann-Roch Theorem

Let $F|K$ be a function field with finite base field. Let $q = |K|$ be the number of elements in K. The residue field for each prime P of F is a finite extension of K, hence in formula (2.2) we have

$$|P| = q^{\deg P} \quad \text{and therefore} \quad |A| = q^{\deg A} \tag{4.1}$$

for all divisors A of the function field. Consequently, after introducing the variable $t = q^{-s}$, F.K. Schmidt's zeta function $\zeta_F(s)$ of (2.2) is transformed into a power series in t:

$$Z_F(t) = \prod_P \frac{1}{1 - t^{\deg P}} = \sum_{n \geq 0} a_n t^n \tag{4.2}$$

with $a_n \in \mathbb{Z}$. It turns out that $Z_F(t)$ is a rational function in the variable t. To see this it is necessary to know more about the coefficients a_n appearing on the right hand side.

In order to simplify notation I will write $Z(t)$ instead of $Z_F(t)$ if it is clear from the context which function field F is discussed at the time. Similarly, I will omit the index F in the notation also for other objects which are connected with F. For instance I am writing W instead of W_F for the canonical class, g for the genus, h

for the class number etc. However, in cases when it is important to indicate which field I am referring to, I shall use the index F or $F|K$.

By definition, a_n in (4.2) is the number of integer divisors of F of degree n. Let C be a divisor class of degree n. The integer divisors in C are counted as follows: Choose a fixed divisor $A_0 \in C$. For any integer divisor A in the same class the quotient $AA_0^{-1} = (f)$ is a principal divisor of some $0 \neq f \in F$, uniquely determined by A up to a nonzero factor in the base field K. (There are $q - 1$ such possible factors). These f (together with the zero element) form a K-vector space $\mathcal{L}(A_0)$, consisting of all multiples of A_0^{-1} in F. Its dimension does not depend on the choice of the divisor $A_0 \in C$ and is denoted by $\dim C$. We conclude that there are

$$a_C = \frac{q^{\dim C} - 1}{q - 1} \tag{4.3}$$

integer divisors A in the class C. Hence

$$a_n = \sum_{\deg C = n} \frac{q^{\dim C} - 1}{q - 1}. \tag{4.4}$$

These numbers appear in the power series expansion (4.2).

In order to obtain information on $\dim C$ the theorem of Riemann-Roch is applied. More precisely I should say the *algebraic analogue* of the Riemann-Roch Theorem. For, neither Bernhard Riemann nor Gustav Roch considered function fields of characteristic p. Their result concerned geometric (or rather topologic) properties of a Riemann surface. Dedekind and Weber had discovered the algebraic nature of the Riemann-Roch Theorem in their pioneering paper of 1882 in Crelle's Journal [DW82] which F.K. Schmidt refers to. But they had worked in characteristic 0 only, with the complex field $K = \mathbb{C}$ as base field. F.K. Schmidt had seen how the Riemann-Roch Theorem can be proved also in characteristic p with arbitrary base field K, and he realized the importance of the theorem for the investigation of the zeta function.

The Riemann-Roch Theorem refers to a certain distinguished divisor class W of the function field $F|K$, called the "canonical class". Sometimes it is called "differential class" since it consists of the divisors of the differentials of $F|K$. Its K-dimension

$$g := \dim W \tag{4.5}$$

is called the "genus" of $F|K$. If the divisors in W are interpreted as divisors of differentials then g can be interpreted as the number of K-linearly independent *integer* differentials, or "differentials of the first kind" as it was called in the classical theory.

This being said, the Riemann-Roch Theorem can now be formulated as follows: For every divisor class C of $F|K$ consider its "dual" WC^{-1}. Then the

Riemann-Roch Theorem establishes a relation between the dimension of C and of its dual:

$$\dim C = \deg C - g + 1 + \dim(WC^{-1}). \tag{4.6}$$

In particular, for $C = W$ it follows

$$\deg W = 2g - 2. \tag{4.7}$$

The canonical class W is uniquely determined by the Riemann-Roch Theorem, and hence (4.6) can be taken as a definition of W. But this does not say anything about the *existence* of a divisor class W satisfying (4.6). F.K. Schmidt [Sch31a] gives a proof which is modeled after that of Dedekind-Weber [DW82] but enhanced with some extra arguments taking into account that the characteristic is a prime number p and, moreover, the base field is not algebraically closed. On the way he obtained the important **Riemann-Hurwitz formula**. This refers to a separating element x of $F|K$. Putting $n = [F : K(x)]$ the Riemann-Hurwitz formula says that

$$2g - 2 = \deg \mathfrak{D}(F|K(x)) - 2n \tag{4.8}$$

where $\mathfrak{D}(F|K(x))$ denotes the different of the separable extension $F|K(x)$.

In later years Hasse included F.K. Schmidt's proof in his book "*Zahlentheorie*" [Has49], and we find the same proof in the later editions of the second volume of van der Waerden's "*Algebra*" [vdW67]. In the meantime quite a number of other proofs have appeared in the literature, one by F.K. Schmidt himself [Sch36], another one by A. Weil [Wei38b], and more. (See also [Che51, Ros52, Roq58]). Today the Riemann-Roch Theorem belongs to the well known prerequisites for the study of function fields, and it is easily accessible in the literature. But this was not yet the case in the 1930s and it is due to F.K. Schmidt to have early seen the importance of the Riemann-Roch Theorem and its relevance for the zeta function.

If $\deg C > \deg W$ then its dual WC^{-1} is of negative degree and hence there are no integer divisors in its class, i.e., $\dim CW^{-1} = 0$. The Riemann-Roch Theorem (4.6) shows that therefore,

$$\dim C = \deg C - g + 1 \qquad \text{if } \deg C > 2g - 2. \tag{4.9}$$

Thus for those C the dimension depends on the degree only. (This is Riemann's part of the Riemann-Roch Theorem). Hence for $n > 2g - 2$ all divisor classes of degree n give the same contribution (4.3) to the series (4.2). Consequently, if h denotes the number of divisor classes of degree n, then

$$a_n = h \cdot \frac{q^{n-g+1} - 1}{q - 1} \qquad \text{if } n > 2g - 2.$$

As already indicated by the notation, the number h of divisor classes of degree n is finite and does not depend on n. Conclusion:

$$Z(t) = \sum_{n \leq 2g-2} a_n t^n + h \cdot \left(\sum_{n \geq 2g-1} \frac{q^{n-g+1} - 1}{q - 1} \right) \cdot t^n \qquad (4.10)$$

The first sum on the right side is a polynomial of degree $\leq 2g - 2$, and the second term a difference of two geometric series, with respect to the powers of qt and of t respectively, starting with the exponent $2g - 1$. (In any case the summation index n in these formulas is assumed to be $n \geq 0$. Hence if $g = 0$ then the first term in (4.10) does not appear and $Z(t) = 1/(1 - qt)(1 - t)$. In the following we will exclude this trivial case, if convenient, and assume $g \geq 1$). We conclude that $Z(t)$ is indeed a *rational function* in t, as announced above already, and that $Z(t)$ has simple poles at $t = q^{-1}$ and $t = 1$.

REMARK 4.3 (DIVISORS OF DEGREE 1)   In the above computations I have tacitly pretended that for every given n there exist divisors in $F|K$ of degree n. This is indeed the case but is not obvious; it was discovered by F.K. Schmidt as a consequence of the finiteness of the base field K. This had raised some surprise among the mathematicians of that time since for infinite base fields this is not always the case. (If $g > 1$ then d is a divisor of $2g - 2$. But if $g = 1$ then, according to Shafarevich [Sha57], there exist function fields $F|\mathbb{Q}$ of genus $g = 1$ for which the smallest positive divisor degree d is arbitrarily large). F.K. Schmidt's proof is analytic in the sense that it uses his analytically defined zeta function in characteristic p. Shortly thereafter, Witt produced an algebraic proof by using suitable arguments from cyclic cohomology [Wit34b]. Both proofs are based on the decomposition law for primes in base field extensions.

REMARK 4.4 (BASE FIELD EXTENSIONS)   Let $F|K$ be a function field with finite base field and K' an algebraic extension of K. Then the field compositum $F' = FK'$ can be regarded as a function field with K' as base field. $F'|K'$ is called a "base field extension" of $F|K$. Already Artin had found that often it is useful to consider base field extensions, see Sect. 3.2.1. F.K. Schmidt in his paper gives a systematic theory of base field extensions. He says:

> "In this way it is possible to avoid the difficulties which, when compared to the classical theory of algebraic functions, are due to the fact that the base field K is not algebraically closed."

The method of base field extensions is used over and over again in the algebraic theory of function fields. Hence it may be useful to list some of their basic properties which F.K. Schmidt had included in his paper. He shows:

- The divisor group of $F|K$ embeds naturally into the divisor group of any base field extension $F'|K'$,
- the degree of a divisor A of $F|K$ does not change if A is considered as a divisor of a base field extension $F'|K'$; similarly for the dimension of A,

- the divisor class group of $F|K$ embeds injectively into the divisor class group of $F'|K'$,
- the canonical class of $F|K$ remains to be the canonical class of $F'|K'$,
- the genus of $F|K$ does not change under base field extensions.

These properties hold not only when the base field K is finite, but more generally for arbitrary separable base field extensions. $K'|K$ may be even a transcendental separable extension (i.e., every finitely generated subextension is separably generated. In such case FK' is defined as the field of quotients of the integral domain $F \otimes_K K'$).

Today a function field $F|K$ is called "conservative" if the genus remains the same in every base field extension, regardless of whether the new base field is separable over K or not. The term "conservative" shows up in Artin's Princeton lectures in the 1950s.

In this book I shall consider base field extensions of a function field only in the case when the function field is conservative, so that I can freely use the above properties.

4.3 F.K. Schmidt's L-polynomial

According to (4.10) $Z(t)$ may be written in the following form:

$$Z(t) = \frac{L(t)}{(1 - qt)(1 - t)} \tag{4.11}$$

where $L(t)$ is a polynomial of degree $2g$. The first two and the last coefficient of $L(t)$ can be explicitly obtained by taking a closer look at (4.10). One finds that

$$L(t) = 1 + (N - q - 1)t + \cdots + q^g t^{2g} . \tag{4.12}$$

Here, I am writing N for the number of primes P of degree 1 of $F|K$, i.e., $N = a_1$ in the notation of (4.2). By writing N instead of a_1 I follow the notation which has become standard in this connection. More precisely I will write N_F when it should be pointed out that this number belongs to the function field F.

In terms of the variable $t = q^{-s}$ the RHp asserts that every root of $L(t)$ lies on the circle $|t| = q^{-\frac{1}{2}}$. It is common to consider the reciprocal polynomial

$$L^*(t) = t^{2g} L(t^{-1}) = t^{2g} + (N - q - 1)t^{2g-1} + \cdots + q^g ; \tag{4.13}$$

its roots are the inverses of the roots of $L(t)$. Hence:

The RHp asserts that all the $2g$ roots of $L^*(t)$ are situated on the circle $|t| = \sqrt{q}$.

While the RHp originally refers to the infinitely many zeros of the analytically defined zeta function, now it appears as a statement about the finitely many roots of a polynomial of degree $2g$. These roots, not necessarily distinct, are algebraic integers, due to the switch from $L(t)$ to $L^\star(t)$.

Let me call $L(t) = L_F(t)$ the "F.K. Schmidt's L-polynomial" of F. Sometimes I shall use this name also for the inverse polynomial $L^*(t)$.

Hasse says in his survey [Has34f] that formula (4.11) had been communicated to him by F.K. Schmidt. But it is not contained in F.K. Schmidt's paper [Sch31a]. Also, I did not find any mention of it in the letters of F.K. Schmidt to Hasse. So I assume that F.K. Schmidt had told Hasse about it at some occasion when they had met. They met several times in the years 1933/34.

4.3.1 Some Comments

On page 41 I have said that F.K. Schmidt had some problems to find the analogue to Kronecker's limit formula for the class number of a function field. Now, for his birationally invariant zeta function he obtained the limit formula from (4.10) and (4.11) as follows:

$$\lim_{t \to 1}(t - 1)Z(t) = -\frac{h}{q - 1} = \frac{L(1)}{1 - q}$$

and therefore simply

$$h = L(1) = L^*(1). \tag{4.14}$$

In the elliptic case, since $L^*(t) = t^2 + (N - q - 1)t + q$ it follows that $h = N$.

When comparing this with Artin's class number formulas (page 23) it should be kept in mind that Artin did not work with F.K. Schmidt's zeta function $\zeta_F(s)$, for the simple reason that this was not yet defined. Artin had studied quadratic function fields $F = K(x, \sqrt{D})$ and he worked exclusively with the truncated zeta function $\zeta_R(s)$ for $R = K[x, \sqrt{D}]$ as explained on page 21. Artin did not know (or care) about the Riemann-Roch Theorem, instead he used the quadratic reciprocity law in $K[x]$, thus keeping close to the analogy to quadratic number fields. His class number formulas (3.8) can easily be transformed into F.K. Schmidt's simple formula (4.14). For, using (3.5) it is seen that in the case of quadratic function fields we have

$$L_R(t) = \begin{cases} L_F(t) & \text{if } \sqrt{D} \text{ is "imaginary" and } \deg D \text{ odd} \\ (1 + t)L_F(t) & \text{if } \sqrt{D} \text{ is "imaginary" and } \deg D \text{ even} \\ (1 - t)L_F(t) & \text{if } \sqrt{D} \text{ is "real"}. \end{cases} \tag{4.15}$$

This corresponds to the fact that $L_R(t)$ may have "'trivial" zeros. In the last two cases there is one trivial zero, namely $t = -1$ or $t = 1$ respectively. In the first case there is no trivial zero.

Moreover it has to be taken into account that when Artin spoke about the "class number" then he meant the number h_R of ideal classes of the Dedekind ring $R = K[x, \sqrt{D}]$. This differs from the class number $h = h_F$ as defined above by a factor $\rho(R)$ called the "regulator" of the Dedekind ring R which is a measure for the size of the fundamental unit in R. This regulator is nontrivial if and only if $\sqrt{D}$ is "real" in the sense of Artin.

By the way, F.K. Schmidt in his thesis [Sch25] defines the regulator $\rho(R)$ of any finitely generated Dedekind ring R of any function field $F|K$ with finite base field K, as follows: Let r denote the number of primes Q_i of $F|K$ which do not lie above R. The group of units $R^\times$ modulo torsion is a free abelian group of $r-1$ generators. Let $\varepsilon_1, \ldots, \varepsilon_{r-1}$ denote a system of fundamental units, i.e., a basis of $R^\times$ modulo torsion. Then he defines

$$\rho(R) := |\det v_i(\varepsilon_j)| \qquad (i, j = 1, 2, \ldots r - 1)$$

where the v_i are the additive valuations of F belonging to the primes Q_i, normalized according to the sum formula for valuations, i.e., $v_i(x) = \deg_{Q_i} v_{Q_i}(x)$. F.K. Schmidt shows that

$$h_R = \rho(R) h_F$$

which generalizes Artin's class number formulas (3.8) on page 23.

For historical reasons I should mention that the existence of fundamental units had been proved already in the year 1903 by Kühne [Küh03]. It appears that F.K. Schmidt did not know that paper. I have already cited another paper by Kühne in connection with the power reciprocity laws in function fields (see page 19). It appears that Kühne's papers were not widely recognized by the mathematical public. This may be due to the fact that they were written in the style of "multiple congruences", and this was not any more understood in later times, after it became common to use the abstract notions of field, ring, ideal etc.

4.4 The Functional Equation

The functional equation describes the behavior of $\zeta_F(s)$ under the substitution $s \to 1 - s$. For the variable $t = q^{-s}$ this means the substitution $t \to q^{-1}t^{-1}$.

The function $t^{-(g-1)}Z(t)$ remains unaltered under the substitution $t \to q^{-1}t^{-1}$. Hence:

$$Z(q^{-1}t^{-1}) = q^{-(g-1)}t^{-(2g-2)}Z(t) . \qquad (4.16)$$

This can be verified by explicitly looking at the expansions (4.2), (4.3) and using the Riemann-Roch Theorem. But I cannot resist to present here an elegant proof by Witt which he has never published. (But it appears, with reference to Witt, in Hasse's report [Has43b] in Italian language).

Witt's proof is based on the Riemann-Roch Theorem (4.6) written in symmetric form:

$$\dim C - \frac{1}{2}\deg C = \dim C' - \frac{1}{2}\deg C' \tag{4.17}$$

where $C' = CW^{-1}$. The degree formula (4.7) can be written in anti-symmetric form:

$$\deg C - (g-1) = -\Big(\deg C' - (g-1)\Big). \tag{4.18}$$

Witt starts by observing the relation

$$\sum_{-\infty < n < \infty} t^n = 0$$

which is to be interpreted as follows: If the sum is cut at any place into two parts, one to the left and one to the right, then the two emerging rational functions in t add up to 0. The relation

$$\sum_C t^{\deg C} = 0 \tag{4.19}$$

is to be interpreted similarly; here and in the following C ranges over all divisor classes of $F|K$, of positive or negative degree.

Now, starting from the expansion (4.2), (4.3) and getting rid of the denominator $q - 1$, Witt applies (4.19) and obtains:

$$(q-1)Z(t) = \sum_C q^{\dim C} \cdot t^{\deg C} \tag{4.20}$$

where he observed that if $\deg C < 0$ then $q^{\dim C} = q^0 = 1$. Introducing the new variable $u = q^{1/2} t$, i.e., $t = q^{-1/2} u$, this can be written as:

$$(q-1)Z(t) = \sum_C q^{\dim C - \frac{1}{2}\deg C} \cdot u^{\deg C}. \tag{4.21}$$

Multiplying with $t^{-(g-1)} = q^{\frac{1}{2}(g-1)} u^{-(g-1)}$:

$$(q-1)t^{-(g-1)}Z(t) = q^{\frac{1}{2}(g-1)} \sum_C q^{\dim C - \frac{1}{2}\deg C} \cdot u^{\deg C - (g-1)}. \tag{4.22}$$

From (4.17) and (4.18) it is seen that on the right side the transformation $u \to u^{-1}$ permutes the terms for C and C', hence the sum remains invariant. By definition, $u \to u^{-1}$ means $t \to q^{-1} t^{-1}$.

4.5 Consequences

The roots of $Z(t)$ are the same as the roots of F.K. Schmidt's polynomial $L(t)$, see (4.11). Their inverses are the roots of $L^*(t)$. Hence as a consequence of the functional Eq. (4.16):

If ω is a root of $L^*(t)$ then $q\omega^{-1}$ is also a root of $L^*(t)$. In other words: The roots of $L^*(t)$ come in pairs ω, ω' (not necessarily distinct) such that $\omega\omega' = q$.

Hence, if it is known that $|\omega| \leq \sqrt{q}$ for all roots ω then one can conclude that $|\omega| = \sqrt{q}$, i.e. the validity of RHp.

The estimate $|\omega| \leq \sqrt{q}$ can be achieved by using Artin's criterion. Artin had established his criterion in his letter to Herglotz (see (3.11) on page 27). He did it for his truncated zeta function $Z_R(t)$ but it also works for F.K. Schmidt's zeta function $Z(t) = Z_F(t)$. Let us explicitly state Artin's criterion for $Z(t)$.

Let $\omega_1, \ldots, \omega_{2g}$ denote the roots of $L^*(t)$. First we observe that if the RHp holds for $F|K$ then, using (4.13) we have:

$$|N - q - 1| = |\omega_1 + \cdots + \omega_{2g}| \leq 2g\sqrt{q}. \tag{4.23}$$

Moreover, the validity of RHp for $F|K$ implies its validity for every base field extension $F_r = FK_r$ where K_r is the extension of K of degree r. This is due to the formula for the respective zeta functions

$$Z_{F_r}(t^r) = \prod_{\varepsilon^r = 1} Z_F(\varepsilon t). \tag{4.24}$$

This is the generalization of Artin's relation (3.9) on page 26, written for the variable $t = q^{-s}$. Observe that K_r has q^r elements, hence for Z_{F_r} the correct variable is $t^r = q^{-rs}$. This explains the appearance of t^r on the left side of (4.24). The formula is locally verified by comparing the contribution of a prime P of F on the right side with the contribution of its prime divisors $Q_1, \ldots, Q_d$ in F_r on the left side. The number d of the Q_i is the greatest common divisor of r and $\deg(P)$. The degree of each Q_i in $F_r|K_r$ is $\deg(P)/d$.

It follows that the zeros of $Z_{F_r}(t)$ are the r-th powers of the zeros of $Z_F(t)$ and therefore, as claimed above, the RHp for $F|K$ implies the RHp for $F_r|K_r$. Thus the estimate (4.23) holds similarly for $F_r|K_r$, i.e.,

$$|N_r - q^r - 1| = |\omega_1^r + \cdots + \omega_{2g}^r| \leq 2g\sqrt{q^r},$$

where N_r is the number of primes of degree 1 of $F_r|K_r$. (Recall that $F_r|K_r$ has the same genus as $F|K$, see Remark 2 on page 46). Artin's criterion is a certain inverse:

Artin's Criterion. *Suppose there exists a constant A independent of r such that*

$$|N_r - q^r - 1| < A\sqrt{q^r} \qquad \text{for all } r \geq 1. \qquad (4.25)$$

Then the RHp holds for the function field $F|K$ (and hence also for every base field extension $F_r|K_r$).

Let us briefly present Artin's proof of it which is contained in his letter of 30 November 1921 to Herglotz (see page 27). Of course Artin worked there in quadratic function fields only and with his truncated zeta function. But his proof is valid also in the general case with F.K. Schmidt's zeta function. Artin writes

$$\log L(t^{-1}) = \log \prod_{1 \leq i \leq 2g} (1 - \omega_i t^{-1})$$

$$= -\sum_{r \geq 1} (\omega_1^r + \cdots + \omega_{2g}^r)\frac{t^{-r}}{r}$$

$$= \sum_{r \geq 1} (N_r - q^r - 1)\frac{t^{-r}}{r}.$$

The terms of this series can be estimated by (4.23) which gives:

$$|\log L(t^{-1})| \leq A \sum_{r \geq 1} \frac{\sqrt{q}^r |t|^{-r}}{r}.$$

The logarithmic series on the right side is convergent for $\sqrt{q}\,|t|^{-1} < 1$. Hence $L(t^{-1})$ has no zero in this region. Thus the ω_i satisfy $\sqrt{q}\,|\omega_i|^{-1} \geq 1$ for $i = 1, \ldots, 2g$. This implies $|\omega_i| \leq \sqrt{q}$, hence $|\omega_i| = \sqrt{q}$ as mentioned above as a consequence of the functional equation. $\qquad\qquad\qquad\qquad\qquad\qquad\qquad\qquad \square$

The relation (4.25) expresses the fact that for $r \to \infty$ the number N_r increases with the order of magnitude q^r with the error term of order $\sqrt{q}^r$. This can be written in the form

$$N_r = q^r + \mathcal{O}(\sqrt{q}^r) \qquad \text{for } r \to \infty \qquad (4.26)$$

where $\mathcal{O}(\cdots)$ is the Landau symbol. Similarly for the number of rational points of an irreducible projective curve Γ defined over K with F as its function field. More precisely, let $N_r(\Gamma)$ be the number of K_r-rational points of Γ. Then $N_r(\Gamma)$ differs from the number N_r of K_r-rational places of F_r by a finite bound, depending on the singularity degree of the curve. If Γ is an affine curve one has also to take into account the finite number of poles. Thus (4.26) is equivalent to

$$N_r(\Gamma) = q^r + \mathcal{O}(\sqrt{q}^{\,r}) \qquad \text{for } r \to \infty. \qquad (4.27)$$

Thus Artin's criterion can also be expressed in geometric language:

Let Γ be an absolutely irreducible curve defined over a finite field K with q elements, and $F = K(\Gamma)$ its function field. For any finite extension $K_r|K$ let $N_r(\Gamma)$ denote the number of K_r-rational points of Γ. Then the condition (4.27) is necessary and sufficient for the validity of the RHp for the zeta function of F.

REMARK Artin's criterion is used in every known proof of the RHp. (Except the proof by Davenport and Hasse for the generalized Fermat fields and the Davenport-Hasse fields. See page 72 and page 74). But it is *not* contained in F.K. Schmidt's papers although, as we have seen, it is an easy consequence of the functional equation. As I have said already, F.K. Schmidt was not primarily interested in the RHp. His aim was to establish the foundations for class field theory for global fields of characteristic p. Moreover F.K. Schmidt did not know Artin's criterion, not even for the case of quadratic function fields. Artin had established this criterion in a letter to Herglotz, written in 1921, but never published it (see page 26 ff). I have found no sign that anyone else except Artin and Herglotz knew this criterion before November of 1932 when Artin met Hasse and told him about it (see next section). Artin's criterion was published in the year 1934 by Hasse [Has34f] in his survey where he collected all known facts about the zeta functions of function fields, including those which he had been told by Artin and F.K. Schmidt. However Hasse's proof of Artin's criterion is different from Artin's proof which I have given above. Hasse's proof is based on Newton's formulas for symmetric polynomials.

Summary

The systematic theory of zeta functions for arbitrary function fields was published by F.K. Schmidt, in generalization of the second part of Artin's thesis. After a preliminary announcement in 1926 F.K. Schmidt decided to build the theory from a birational point of view, in the spirit of the classical work of Dedekind-Weber of 1882. The basic notions are those of "prime divisor" (or briefly "prime") of a function field, and of "divisor". F.K. Schmidt proved the Riemann-Roch Theorem for divisors of algebraic function fields with perfect base fields. If the base field is finite then he exhibited a close relationship between the Riemann-Roch Theorem and the properties of the zeta function of the function field, in particular the functional equation. This is to be regarded as his main achievement. The zeta function is essentially, up to a trivial factor without zeros, a polynomial $L(t)$ of degree $2g$ where g is the genus of the field. A direct consequence of the functional equation is Artin's criterion for the validity of the RHp. But this is not contained in F.K. Schmidt's paper. It was published in the year 1934 by Hasse after Artin had orally informed him about it.

The main aim of F.K. Schmidt was not the RHp but the establishment of class field theory for function fields over finite base fields. (He did not fully reach this goal, however.) F.K. Schmidt's paper was used as his thesis for Habilitation 1927 in Erlangen, and it was published 1931 in the Mathematische Zeitschrift [Sch31a], the same journal where Artin's thesis had appeared.

There is evidence that Artin in his lectures in the 1920s also developed the algebraic theory of function fields, and perhaps the analytic theory too. But this had not appeared in print at that time.

Chapter 5
Enter Hasse

© Springer Nature Switzerland AG 2018
P. Roquette, *The Riemann Hypothesis in Characteristic* p
in Historical Perspective, Lecture Notes in Mathematics 2222,
https://doi.org/10.1007/978-3-319-99067-5_5

Helmut Hasse (1898–1979) was of the same age as Artin.[1] In 1917 during his
military service he got permission to study at the University of Kiel with Toeplitz.
After the war he moved to Göttingen where he studied mainly with Hecke. When
the latter left Göttingen in 1920, Hasse became interested in the theory of p-adic
numbers and went to Marburg to study with Kurt Hensel. He regarded Hensel as his
"first" academic teacher, the "second" being Erich Hecke. (That's what he answered
when once I had asked him about it.)

In May 1921 Hasse received his Ph.D. in Marburg. In his thesis he formulated
his "Local-Global Principle" for quadratic forms over the rational number field $\mathbb{Q}$.
His papers on quadratic forms culminated 1924 in the Local-Global Principle for
quadratic forms over an arbitrary algebraic number field [Has24]. In the year 1922
Hasse obtained a position as *Privatdozent* at the University of Kiel. The towns of
Kiel and Hamburg are not far from each other, and the mathematicians of Kiel kept
close contacts to their colleagues in Hamburg. There, Hasse and Artin frequently
met in Hecke's seminar, and a long lasting friendly relationship began, documented
in a number of letters between Artin and Hasse, mostly on problems of class field
theory [FLR14].

In the year 1925 Hasse accepted a professorship at the University of Halle. In
the summer semester 1930 he changed to Marburg, as the successor of his academic
teacher and "fatherly friend" (*väterlicher Freund*) Kurt Hensel.

Of course Hasse was familiar with the content of Artin's thesis. In fact he
had reviewed it in the *Jahrbuch der Fortschritte der Mathematik*. Although he
briefly mentioned the RHp in his review I have found no evidence that Hasse was
particularly interested in the RHp at that time. But this suddenly changed in 1932.
One can precisely determine when and how Hasse became seriously interested in
the RHp. This happened in one of the last days of November 1932. On this day
Hasse arrived in Hamburg by train from Kiel where he had given a colloquium talk
at the University, on invitation from Fraenkel.

Hasse had known Fraenkel since many years; their correspondence shows their
friendly respect towards each other.[2] Fraenkel was full professor at the University
of Kiel since 1928 . There he tried whatever he could do to have Hasse (who
was in Halle at that time) back to Kiel University, but this was in vain since the
Prussian ministry of education wanted Hasse to go to Marburg as the successor of
Hensel. In fact, as I have said already Hasse went to Marburg in 1930. In September
of 1932 Fraenkel met Hasse again at the International Congress of Mathematicians
in Zürich, and he invited Hasse for a colloquium talk in Kiel. Hasse accepted; the

---

[1] In their joint paper [AH25] it is said that the write-up of the paper was done by the *younger* of
the two. Since both were born in the same year it takes a precise knowledge of their birth dates to
decide who indeed was the younger one. Fact is that Hasse was 175 days younger than Artin.

[2] The letters from Hasse to Fraenkel are kept at the archives of Hebrew University in Jerusalem. I
am indebted to Moshe Jarden for helping me to find them. The letters in the other direction are in
the Hasse *Nachlass* in Göttingen.

date was finally fixed for the end of November of 1932. (Fraenkel had invited Hasse to stay in his (Fraenkel's) home while in Kiel for the colloquium.)

Originally Hasse had proposed to talk in Kiel on his recent results on simple algebras over number fields and their role in class field theory—this had also been the topic of his lecture at the ICM in Zürich. In his *Nachlass* I have found a complete script for this proposed talk. But some time between September and November Hasse apparently had become interested in another topic, namely diophantine congruences. In his *Nachlass* there is a second script for a talk in Kiel with the title:

> "*On the asymptotic behavior of numbers of solutions of congruences modulo* p."

Hasse had recently become interested in this topic through his friend Harold Davenport. It seems not likely that Hasse gave two talks in the colloquium at Kiel. I believe that he talked about the second topic only, like he did in Hamburg some days later.

As the towns of Kiel and Hamburg are situated not far from each other it seems natural that Hasse, on the way back from Kiel, took a stop in Hamburg "*with the only purpose to be together with the Artins*" as he wrote to Davenport. But somehow Artin had been able to induce Hasse to present his Kiel lecture a second time in Hamburg. This lecture had the same title as that in Kiel which I have quoted above. There is no hint in the title nor in Hasse's own lecture notes of a connection to function fields and the RHp.

But such a connection was opened by Artin in the discussion with Hasse after his colloquium talk. Artin informed Hasse about his results on the RHp which he had written to Herglotz but never published (see Sect. 3.2). In particular he told Hasse about his criterion for the validity of the RHp (see page 52). In view of these results the Hasse-Davenport problem mutated into the RHp—and for the next years Hasse was almost completely absorbed in its proof.

But what was Hasse's original problem on diophantine congruences and how did he become interested in it? This is another side of our story. Hasse's younger friend Davenport has played a decisive role in this. One can safely say that without his friendship with Davenport, Hasse would not have taken up his work on the RHp.

Chapter 6
Diophantine Congruences

6.1 Davenport

© Springer Nature Switzerland AG 2018
P. Roquette, *The Riemann Hypothesis in Characteristic* p
in Historical Perspective, Lecture Notes in Mathematics 2222,
https://doi.org/10.1007/978-3-319-99067-5_6

Harold Davenport had been introduced to Hasse in 1930 by Louis J. Mordell. This came about as follows:

On 25 November 1930 Hasse wrote a letter to Mordell who at that time held a professorship at the University in Manchester. They knew each other since several years; they had exchanged reprints and letters. Mordell, who liked to travel, had visited Hasse occasionally; the last visit had been on 16 July 1930 when Mordell gave a colloquium talk in Marburg on the invitation of Hasse. The above mentioned letter contained, on the request of Mordell, Hasse's opinion on Wedderburn who was considered for election to the Royal Society. Hasse had written his opinion in German because he felt not to be sufficiently proficient in the English language. He added a note to his letter, in a rather quaint[1] English, which reads as follows:

> "... In order to have further occasion for applying and enriching my knowledges I would much like to get a young English fellow at home. It would be very kind of you, if you could send me one of your students during next summer term (April-July). We would invite that student to dwell and eat with us. He would be obliged to speak English with us at any time we are together (at breakfast, dinner, tea, lunch etc.)... From my point of view it would be best, if he were student of pure mathematics out of an advanced course of yours... I would much like to hear from you, whether you know a clever and handsome fellow for this purpose."

Thus Hasse wished to upgrade his English. At those times, English had not yet become the *lingua franca* for science and not, in particular, for mathematics.

Already 2 days after Hasse had sent the letter, on 27 November 1930, Mordell replied to him as follows:

> "... I can suggest the very person you want to go to Marburg. Mr. Harold Davenport, Trinity College, Cambridge. He was formerly one of our students, the best we have had for many years. He is now doing research, and lately he has proved some such result as $\sum_{n=0}^{p-1} \left(\frac{n^4 + an^2 + bn + c}{p} \right) = \mathcal{O}(p^{3/4})$ where the left hand ( ) is the symbol of quadratic reciprocity ... He is interested in certain aspects of number theory and I believe he would be free to go. I have written to him and asked him to write direct to you ..."

Davenport needed not much time to think it over. On 30 November 1930 he wrote to Mordell thanking him for passing Hasse's request on to him. He regarded it as a great compliment, especially, he wrote, in view of the phrase "handsome fellow" which appeared in Hasse's letter. And he was very interested in Prof. Hasse's scheme. On 7 December 1930 Davenport wrote to Hasse from Cambridge:

> "Dear Prof. Hasse,
> Prof. Mordell has told me of your letter to him, in which you say that you would like to know of an advanced English student of pure mathematics,

---

[1] This is the expression which Davenport had used in a letter to Mordell.

whom you could invite to Marburg next summer term. May I offer you my services?

I used to be a student of Mordell's at Manchester, but for the last three years I have been studying here. I am particularly interested in the analytical theory of numbers – Gitterpunktprobleme, ζ-function, etc. Are you interested in these subjects, or is there anyone else at Marburg who is? So far I have only written two short papers, which will appear soon in the Journal of the London Mathematical Society; one on the distribution of quadratic residues mod (p), the other on Dirichlet's L-functions.

I am 23 years old, and not at all 'handsome' (as you required in your letter). Also I do not swim or drink beer – and I understand that these are the principal recreations in Germany..."

Hasse seems not to have minded these "shortcomings" with which Davenport had advertised himself, and he sent Davenport a definite invitation.

So in the next summer (1931) Davenport stayed as "language teacher" with the Hasses in Marburg. There developed a longtime friendship between the Hasse family and the younger Davenport—including several mutual visits during the next years. Certainly this had an effect on Hasse's proficiency in English, but in addition Davenport succeeded to raise Hasse's interest in English history, English literature and quite generally in everything which was considered as "typically English". Hasse kept this interest throughout his life. On the other hand, Davenport also profited from this contact; later on he was fluent in German. Let us cite from a letter of Davenport to Mordell in September 1931, from the German town of Bad Elster where he was staying with Hasse during the annual meeting of the German Mathematical Society.

"The Hasses and I have been on a motor tour during the past 12 days, in which we have visited the Black Forest, Switzerland, the Italian Lakes, and Tyrol, with very much pleasure and edification. Hasse is now taking an active part in the D. M. V. congress here, I a more passive part. I can never be sufficiently grateful to you for passing on Hasse's invitation to me: I have had an excellent time in Marburg."

By the way, the car of that motor tour was Davenport's. At that time Hasse did not own a car.

6.2 The Challenge

Of course, the conversation between the two was not confined to English language and literature but it soon included mathematics. It seems quite natural that one of the first questions of Hasse to his younger colleague was about Davenport's result which Mordell had mentioned in his letter. (See page 60.) On first sight this seems to be a rather special technical lemma. But since Mordell had mentioned this explicitly, Hasse wanted to know more details.

Consider the curve Γ defined over $\mathbb{Z}$, given by

$$\Gamma : y^2 = x^4 + ax^2 + bx + c$$

with rational integers a, b, c. Let $p > 2$ be a prime. In order to count the rational points of Γ modulo p one has to count those x mod p for which the right hand side is a quadratic residue mod p. Observe that

$$1 + \left(\frac{z}{p}\right) = \begin{cases} 2 & \text{if } z \text{ is a quadratic residue mod } p \\ 0 & \text{if } z \text{ is a quadratic non-residue mod } p \\ 1 & \text{if } z \equiv 0 \text{ mod } p \end{cases} \tag{6.1}$$

where the brackets denote the symbol for quadratic residues modulo p as in Mordell's letter. In any of these cases, $1 + (\frac{z}{p})$ is the number of solutions y modulo p of the congruence $y^2 \equiv z$ mod p. We see that the number of solutions (x, y) mod p of the congruence

$$y^2 \equiv x^4 + ax^2 + bx + c \quad \text{mod } p \tag{6.2}$$

is given by the sum of the p terms $1 + \left(\frac{x^4+ax^2+bx+c}{p}\right)$ when x ranges over $\mathbb{Z}$ modulo p. Hence this number of solutions is

$$p + \sum_{x \bmod p} \left(\frac{x^4 + ax^2 + bx + c}{p}\right).$$

Here appears the sum mentioned in Mordell's letter (see page 60). Accordingly Davenport's result can be expressed as follows:

If p is large then the number of solutions mod p of the congruence (6.2) *is about p with the error term of order of magnitude $\mathcal{O}(p^{3/4})$ for $p \to \infty$.*

Here, the symbol $\mathcal{O}(\cdots)$ is, like in Mordell's letter, the so-called Landau symbol; the above statement means

$$\left| \sum_{x=0}^{p-1} \left(\frac{x^4 + ax^2 + bx + c}{p}\right) \right| < C \cdot p^{\frac{3}{4}}$$

with a constant C independent of p. The really important achievement of Davenport was that he could get the estimate of the error term with the exponent $\frac{3}{4}$. This was much better than all earlier known estimates. Davenport's method for this was new and consisted of an ingenious manipulation of the sum to be estimated. See [Dav31]. (In that paper Davenport obtained the above estimate while studying the occurrence of consecutive quadratic residues modulo p.) However, heuristically the exponent $\frac{1}{2}$

was expected as the proper estimate. Thus Davenport's result, although the best estimate known at the time, was not considered to be the final one.

Looking at the congruence (6.2) we are reminded of Artin's thesis where already in 1921 such congruences were investigated. In fact, the congruence (6.2) defines a quadratic function field $F = \mathbb{F}_p(x, y)$ over the prime field $\mathbb{F}_p$ of characteristic p. The solutions of that congruence correspond to the prime ideals of degree 1 in the ring $R = \mathbb{F}_p[x, y]$ which appear in Artin's zeta function $\zeta_R(s)$ of F. The suspected exponent $\frac{1}{2}$ in the estimate of the error term would have been a consequence of the RHp for these function fields (see (3.11) on page 27).

But Artin was not mentioned in Davenport's paper [Dav31]. At the time when Davenport had written this paper in 1930 he did not yet know Artin's thesis. He learned about it from Hasse during his visit in Marburg, as reported by Halberstam in [Dav77].

Inspired by Davenport's result, Mordell too became interested in the question, and he studied a number of other instances where one could obtain similar estimates. Mordell's paper appeared in 1933 but he had obtained the results in 1931 already . For, on 8 November 1931 he wrote to Hasse:

"...*during the last three weeks I became very interested in Davenport's note on the distribution of quadratic residues and I could not do anything else. I have only within the last few days proved that the number of solutions of* $y^2 \equiv ax^3 + bx^2 + cx + d \bmod p$ *is* $p + \mathcal{O}(p^{3/4})$ *& more generally when* y^2 *is replaced by* y^m *except in one trivial case. Davenport has also found the theorem & proof of a different case about the same time. If I remember any German, I might speak on this to your students etc. as the method is very elementary.*"

The last sentence refers to Mordell's future visit to Marburg, planned for early 1932. One month later, on 14 December 1931 Mordell wrote again:

"...*You may also be interested in knowing that I have made further progress with congruences. The cubic congruence* $f(x, y) \equiv 0$ *has in general* $p + \mathcal{O}(p^{2/3})$ *solutions. Also* $y^m \equiv a_1 x^n + \cdots + a_{n+1}$ *has in general* $p + \mathcal{O}(p^{\gamma(m,n)})$ *solutions where*

$$\gamma(m, n) = 2/3 \quad \text{if} \quad n = 4, \ m = 2$$
$$= 7/8 \quad \text{if} \quad n = 6, \ m = 2$$
$$= 5/6 \quad \text{if} \quad n = 4, \ m = 4$$
$$= 3/4 \quad \text{if} \quad n = 3, \ (m = 2 \text{ included above})$$
$$= 1/2 \quad \text{if} \quad n = 3, \ m = 3 .$$

Davenport has also found results of this kind; and I saw him three days ago..."

As to Davenport's results, he had obtained in addition:

$$\gamma(4,4) = \frac{2}{3} \,,\, \gamma(m,4) = \frac{5}{6} \,,\, \gamma(5,5) = \frac{5}{6} \,,\, \gamma(3,6) = \gamma(6,6) = \frac{7}{8} \,,\, \gamma(2,7) = \frac{19}{20}$$

in his papers [Dav31] and [Dav33].

I have found no indication that Hasse had developed a special interest in problems of this sort before he was confronted with these results by Mordell and Davenport. But now, looking at these and their proofs which often required some nontrivial computational skill, he became interested. Nevertheless he was not much impressed by the methods used, and not by the general attitude of the authors towards these problems.

Hasse tended to think about diophantine congruences not as problems *per se*, but as manifestations of mathematical structures. Thus, he said to Davenport, the results so far are obtained through clever computations only, manipulating and estimating algebraic and analytic expressions. Hasse acknowledged that the methods used may be non-trivial but they did not seem to him adequate since they lead to many different exponents for the remainder term in so many special cases, whereas in every case the exponent $\frac{1}{2}$ was expected. Perhaps it would be possible to reduce some of the exponents a little further by refining those methods. But instead of "reducing exponents" the proper thing to do would be to find out the structure behind this.

(The terminology "reducing exponents" seems to have been established between Hasse and Davenport, somewhat ironically on the side of Hasse (you should do better than reducing exponents!) and in a sense provocative on the side of young Davenport (I can at least reduce exponents, and what can you do with your abstract methods?). In a letter of 25 February 1932 Davenport had written "*I haven't reduced any exponents recently, I regret to say.*")

Davenport may have said that he does not believe that those abstract methods could do better than the very explicit computations which he and Mordell had used. After all, what only counts are the results and not the methods. And since Hasse still insisted on his view, Davenport challenged him to solve the problem with his abstract structural methods.

The above description of a possible dispute between Hasse and Davenport is not purely fictional. Hasse used to tell us this story along these lines when asked about it. At the other side, Davenport told the story in the same spirit to Mordell to whom he was quite close. Mordell reports in [Mor71]:

> "*Davenport was staying with Hasse at Marburg in the earlier thirties and challenged him to find a concrete illustration of abstract algebra. This led Hasse to his theory of elliptic function fields...*"

Hasse accepted Davenport's challenge. In September 1932, after the Zürich congress of the International Mathematical Union where Hasse reported on his results on simple algebras, he found time to start working on the problem.

As a first step, he tried to generalize the problem, replacing the integers $\mathbb{Z}$ and its prime numbers p by the integers of an arbitrary algebraic number field and its prime ideals $\mathfrak{P}$. Thus, given an absolutely irreducible polynomial $f(x, y)$ with integer coefficients in that number field, the problem is to count the solutions of $f(x, y) \equiv 0 \bmod \mathfrak{P}$ in that number field.

We see that Hasse did what mathematicians often do: in analyzing a problem he tried to generalize it in order to find out which properties are essential for the problem and which are not. Hasse went through the special examples of Davenport and Mordell. He was able to get the same exponents γ of the above list in all those cases, but now for congruences modulo prime ideals in number fields. He had found out about it while travelling from Kiel to Hamburg by train. (As I have reported earlier in Chap. 5, Hasse had given colloquium talks in Kiel and Hamburg in November of 1932.) In a letter to Davenport dated 7 December 1932 Hasse wrote:

"My lectures found much interest with the Hamburg and Kiel mathematicians. In Hamburg, I was able to produce a couple of new results, which I had found during my journey back from Kiel in a Personenzug."[2]

But working with congruences modulo prime ideals in number fields amounts to the same as working with equations in finite fields. And it was Artin who, after Hasse's talk in Hamburg, informed him that these results have some bearing on the location of the roots of the zeta function of the respective function fields in characteristic p. Well, this did not yet lead to the proof of the RHp for these function fields since the exponent γ appearing in Mordell's and Davenport's results was not always $\frac{1}{2}$. But Artin's criterion and proof works also for arbitrary exponents γ with $\frac{1}{2} \leq \gamma < 1$, leading to the conclusion that the roots ω of F.K. Schmidt's L^*-polynomial satisfy at least $|\omega| \leq q^\gamma$ (compare with Sect. 4.5).

As said above already, in all these cases the exponent $\gamma = \frac{1}{2}$ was expected. This could now be interpreted that in all these cases the validity of RHp was expected.

Thus Hasse, after his discussion with Artin, had now found the algebraic structure behind the various estimates of Davenport and Mordell, namely global fields in characteristic p and the behavior of their zeta functions.

It appears that Hasse at that time was not yet fully convinced that the RHp would be true in all cases. I am indebted to the late Professor S. Iyanaga, who had been present at Hasse's Hamburg talk in 1932, for informing me that in the discussion Hasse was still somewhat hesitating. In contrast, Artin showed himself quite certain that the RHp will hold generally.

It is a common observation in the history of mathematics that a problem can be more easily solved if it can be put into a structural framework which seems "adequate" to it—at least in the eyes of those people who work on that problem. In any case, after his Hamburg visit with Artin, Hasse found that the framework of function fields and their zeta functions was adequate to the Davenport-Mordell problem.

Let Hasse himself explain his new vision in his own words, to be found in his *Zentralblatt* review of Mordell's paper [Mor33]. The results of this paper had

---

[2]At that time in the language of German Railways a "Personenzug" meant a slow train. It took about 50 minutes from Kiel to Hamburg. Fast trains were called "D-Zug".

already been announced by Mordell in his letter to Hasse on 14 December 1931 (see page 63.) The following text is an excerpt from Hasse's review.

> *"The paper is concerned with special cases of the following general problem: Let $f(x, y)$ be a polynomial with integer coefficients which is absolutely irreducible over the finite field $\mathbb{F}_p$ of p elements, and N the number of solutions of $f(x, y) = 0$ in $\mathbb{F}_p$. One should find an estimate of the form*

$$ (A): \qquad |N - p| \leq C p^\gamma $$

> *where the exponent $\gamma < 1$ is as small as possible, and C is a positive constant. Both γ and C should not depend on p, and also not on the special choice of the coefficients of f, but only on the algebraic invariants of the function field F defined by the equation $f(x, y) = 0$ over $\mathbb{F}_p$.*
>
> *I would like to remark in advance that the final solution of this general problem is closely related to the analogue of the Riemann hypothesis for F. K. Schmidt's zeta function $\zeta_F(s)$ for F. If the infinite solutions are correctly included into the count then N becomes the number of prime divisors of degree 1 of F and the theory of $\zeta_F(s)$ shows that γ can be chosen as the maximal real part θ of the zeros of $\zeta_F(s)$. In addition, one can choose $C = 2g$ where g is the genus of F. It is known that $\theta < 1$ but a bound which is independent of p is not yet known. The analogue of the Riemann hypothesis, $\theta = \frac{1}{2}$, would imply that one could choose $\gamma = \frac{1}{2}$.*
>
> *Conversely, the statement (A) for F and for all constant field extensions of F (where on both sides p is to be replaced by the number $q = p^r$ of elements in the field of constants, and γ, C are independent of r too) would imply that $\theta \leq \gamma$, hence for $\gamma = \frac{1}{2}$ the analogue of the Riemann hypothesis for F would follow. – The author explains this connection in the hyperelliptic cases $f(x, y) = y^2 - f(x)$ only, with the special congruence zeta functions of Artin."*

Only after this introduction Hasse proceeds to review the results of Mordell's paper in some more detail; these are essentially the same as Mordell had stated in his letter to Hasse of 14 December 1931, where he obtained the values $\gamma = \frac{2}{3}, \frac{7}{8}, \frac{5}{6}, \frac{3}{4}, \frac{1}{2}$ in various situations; see page 63.

The above review text shows clearly Hasse's new viewpoint. Whereas Mordell regarded the theory of the zeta function as a means to obtain good estimates of the form (A), Hasse now proposes to look in the other direction. Namely, any estimate of the form (A) would lead to a result about the real parts of the zeros of the zeta function, *provided that the estimate (A) can be proved over all finite field extensions of $\mathbb{F}_p$ too.* In this spirit Hasse added a last paragraph to his review, confirming what he had found in the train from Kiel to Hamburg and reported in his letter to Davenport:

> *"I would like to add that the results of the author can be transferred almost word for word to the case of an arbitrary finite field K instead of $\mathbb{F}_p$ as field*

of coefficients. Hence, as said at the beginning, they lead to a bound of the maximal real part of zeros θ by the respective γ and, if γ = $\frac{1}{2}$, to the Riemann hypothesis for the respective zeta function of F. K. Schmidt."

Three months after his visit to Hamburg, at the end of February 1933, Hasse succeeded to prove the Riemann hypothesis in the case of elliptic function fields. I shall report about it in Chap. 7.

6.3 The Davenport-Hasse Paper

When in a conversation the work of Hasse on the RHp comes up then, in my experience, most people connect his name with *elliptic* function fields (or curves) only. It seems to be less known that he also obtained the RHp for other classes of fields, namely the generalized Fermat fields and the Davenport-Hasse fields. And this happened even before the proof for elliptic fields was complete. The immediate cause for this was a letter from Davenport.

6.3.1 Davenport's Letter and Generalized Fermat Fields

As said above, Davenport stayed with the Hasses in Marburg in the summer semester of 1931. Already in January of 1932 Davenport again visited Hasse. As Hasse wrote to Mordell on 20 November 1931:

"We have invited Davenport for the second half of his Christmas vacation – or rather he has invited himself with our readily given consent..."

Apparently on that occasion Davenport informed Hasse about his latest result which concerned the number of solutions of congruences of the form

$$ax^m + by^n + c \equiv 0 \bmod p \qquad \text{(if } a, b, c \not\equiv 0 \bmod p).$$

Hasse seems to have asked him for details, for as soon as Davenport was back in England he sent Hasse the full proof. The letter is not dated but Hasse wrote "Jan 1932" on the margin. It turned out that for this kind of congruence Davenport had been able to obtain the best estimate, i.e., with the exponent $\gamma = \frac{1}{2}$.

Davenport's proof seems to have caught Hasse's interest, for right away he copied it into his mathematical diary [LR12]. Davenport's letter is the nucleus of the joint paper of Davenport and Hasse [DH34]. As the proof is short and beautiful let us read Davenport's letter:

"My dear Helmut, I promised to send you my treatment of the congruence

$$ax^m + by^n + c \equiv 0 \pmod{p}. \tag{6.3}$$

Let $\chi_1, \ldots, \chi_{m-1}$ be the non-principal characters for which $\chi^m = 1$, the principal character. It is easily seen that

$$1 + \chi_1(t) + \cdots + \chi_{m-1}(t) \tag{6.4}$$

is precisely the number of solutions of $x^m \equiv t$. Hence the number of solutions of (6.3) is

$$N = \sum_t \left\{ 1 + \chi_1(t) + \cdots + \chi_{m-1}(t) \right\} \times$$

$$\left\{ 1 + \psi_1\left(-\tfrac{at+c}{b}\right) + \cdots + \psi_{n-1}\left(-\tfrac{at+c}{b}\right) \right\}$$

where $\psi_1, \ldots, \psi_{n-1}$ are the non-principal characters for which $\psi^n = 1$. Hence

$$N = p + \sum_{r=1}^{m-1}\sum_{s=1}^{n-1}\sum_t \chi_r(t)\psi_s(-\frac{at+c}{b}).$$

The sums in t can be easily expressed in terms of generalized Gauss sums

$$\tau(\chi) = \sum_v \chi(v)e(v), \qquad e(x) = e^{\frac{2\pi i x}{p}}. \tag{6.5}$$

These have the property $\overline{\chi}(u)\,\tau(\chi) = \sum_v \chi(v)e(uv)$. Hence

$$\sum_t \chi(t)\psi(at+c) = \frac{1}{\tau(\overline{\psi})}\sum_{t,v}\chi(t)e((at+c)v)\overline{\psi}(v)$$

$$= \frac{\tau(\chi)}{\tau(\overline{\psi})}\sum_v \overline{\chi}(av)\,\overline{\psi}(v)\,e(cv)$$

$$= \frac{\tau(\chi)\,\tau(\overline{\chi}\,\overline{\psi})}{\tau(\overline{\psi})}\,\overline{\chi}(a)\,\chi\psi(c).$$

Therefore

$$N = p + \sum_{r=1}^{m-1}\sum_{s=1}^{n-1}\frac{\tau(\chi_r)\tau(\overline{\chi}_r\overline{\psi}_s)}{\tau(\overline{\psi}_s)}\,\chi_r\left(\frac{c}{a}\right)\psi_s\left(-\frac{c}{b}\right) \tag{6.6}$$

$$= p + \vartheta\sqrt{p}\,(m-1)(n-1),\ \left(|\tau| = \sqrt{p},\ |\vartheta| \leq 1\right)$$

$$> 0 \ \text{ if } \ p > (m-1)^2(n-1)^2.$$

Quite trivial ! . . ."

In the above letter Davenport has in mind the characters χ, ψ of the multiplicative group $\mathbb{F}_p^\times$. He does not explicitly mention that m and n are supposed to divide $p - 1$, which is natural in this situation. (For otherwise, m, n could be replaced by their greatest common divisor with $p - 1$ without changing the number N of solutions.) If $t \equiv 0 \bmod p$ he puts $\chi_r(t) = 0$ and similarly $\psi_s(t) = 0$. Moreover, a, b, c are supposed to be $\not\equiv 0 \bmod p$. The group theoretical character relation (6.4) generalizes the statement (6.1) in the preceding section.

Davenport's computation yields

$$|N - p| \leq C \cdot \sqrt{p} \qquad \text{with } C = (m - 1)(n - 1). \tag{6.7}$$

Such estimate is also contained in a paper by Mordell but with another constant C and not using Gauss sums; see [Mor33]. Mordell had visited Hasse in January 1932 and on that occasion given him a copy of his manuscript. Hence Hasse knew about it. Mordell's proof is "purely elementary". But Hasse did not copy Mordell's proof into his diary. Evidently he preferred Davenport's who used Gauss sums.

It seems likely that the estimate (6.7) was among those which Hasse could generalize, later in the year, to the case of arbitrary finite fields instead of just the prime fields $\mathbb{F}_p$. (See page 65.) Namely it turned out that Davenport's computation works in the same way over any finite field $K = \mathbb{F}_q$, provided m, n divide $q - 1$. In that case χ, ψ range over the nontrivial characters of the multiplicative group $K^\times$ with orders dividing m, n respectively. The exponential $e(x)$ appearing in Davenport's definition of the Gauss sum has now to be defined as

$$e(x) = e^{\frac{2\pi i S(x)}{p}} \tag{6.8}$$

where $x \in K$ and $S : K \to \mathbb{F}_p$ denotes the trace function ("*Spur*" in German). Hence we can conclude:

> Let K be any finite field and q the number of its elements. Assume m and n divide $q - 1$. Then the number N of solutions in K of the equation
>
> $$ax^m + by^n + c = 0 \quad \text{with } a, b, c \in K^\times \tag{6.9}$$
>
> satisfies the estimate (6.7) with p replaced by q.

But then the same holds also over every finite extension $K_r = \mathbb{F}_{q^r}$ and so we have (6.7) also for q^r. It follows from Artin's criterion (see page 52):

> The RHp holds for any function field $F = K(x, y)$ over K which can be generated by an equation of the form (6.9).

Observe that here it is not necessary any more to assume that m, n divide $q - 1$. For, if this would not be the case then one could replace K by any finite extension of K for which this is the case. We know from Artin's result on base field extensions (Sect. 3.2.1) that the RHp for that extension implies the RHp for $F|K$.

Function fields defined by (6.9) are called "**generalized Fermat fields**".

Thus already in 1932 Hasse had proved the RHp for a large class of function fields of genus $g > 1$. He used Davenport's method of estimating by Gauss sums, generalized it to arbitrary finite base fields and then applied Artin's criterion. Davenport called this the "GF-method" since it provided a proof by extending the base field, which was $\mathbb{F}_p$ in Davenport's papers, to any Galois field (=GF). (At that time finite fields were also called "Galois fields", at least in the English mathematical literature.)

REMARK Consider the double sum in (6.6). Since $\tau(1) = 0$ it suffices to sum over those pairs r, s for which $\chi_r\psi_s \neq 1$. Let d denote the greatest common divisor of m and n. There are $d - 1$ terms in the double sum (6.6) with $\chi_r\psi_s = 1$ and which therefore can be omitted in (6.6). Thus Davenport's estimate (6.7) can be sharpened to $C = (m - 1)(n - 1) - (d - 1)$. But the genus g of the function field is given by

$$g = \frac{1}{2}\Big((m - 1)(n - 1) - (d - 1)\Big). \qquad (6.10)$$

Thus the Davenport-Hasse computation yields

$$|N - q| \leq 2g\sqrt{q}. \qquad (6.11)$$

Observe that here N stands for the number of solutions of (6.9). If counting the number of prime divisors of degree 1 of the function field then one has also to include the pole P_∞ of x and y, and with this new N the above inequality reads

$$|N - q - 1| \leq 2g\sqrt{q}, \qquad (6.12)$$

which is precisely what is expected for the RHp.

6.3.2 Gauss Sums

Hasse and Davenport agreed to have a joint paper containing the proof of the RHp for generalized Fermat function fields. That paper appeared in 1934 in Crelle's Journal; see [DH34] Sometimes this paper is erroneously cited as having appeared in 1935. But the part of volume 172 which contained the Davenport-Hasse paper had appeared in 1934 already. By the way, the results of the paper had already been announced by Hasse in September of 1933 at the annual meeting of the German Mathematical Society (DMV) [Has34c].

The Davenport-Hasse paper contains much more than just the RHp. While the RHp is concerned only with the absolute value $|\omega|$ of the roots ω of F.K. Schmidt's L-polynomial of $F|K$, the Davenport-Hasse paper deals also with the question of identifying those roots completely.

In this spirit Davenport had asked Hasse in a letter of 17 March 1933:

"What do you think the form of the ordinates of the zeros of Artin's ζ–function will be?"

In terms of F.K. Schmidt's polynomial $L^*(t)$ this is to be translated into the question about the angles of its roots while the RHp is concerned with their absolute value. But Hasse wanted more. The roots ω of $L^*(t)$ are algebraic integers, accordingly he wished to characterize them not only as complex numbers but through their *arithmetic* properties, i.e., prime decomposition and congruences. In the general case, for arbitrary function fields, this is a difficult question and probably not possible to solve in a meaningful way. But in the case of the generalized Fermat function fields, Hasse's dream came true. (Also in the case of Davenport-Hasse fields, see Sect. 6.3.3, and of elliptic fields, see Chap. 7.)

For simplicity, only the case without parameters a, b, c was considered since this can be achieved by a suitable finite extension of the base field. So the Eq. (6.9) becomes

$$x^m + y^n = 1. \tag{6.13}$$

Theorem *Let K be a finite field with q elements and $F = K(x, y)$ the generalized Fermat function field with the defining relation (6.13). It is assumed that m and n divide $q - 1$. Then the zeros of F.K. Schmidt's polynomial $L^*(t)$, i.e., the inverses of the zeros of the zeta function $Z(t)$, are given by the so-called Jacobi sums*

$$\omega_{\chi, \psi} := - \sum_{a+b=1} \chi(a)\psi(b) \tag{6.14}$$

where χ , $\psi \neq 1$ range over the nontrivial characters of $K^\times$ of order dividing m and n respectively, with the specification that $\chi\psi \neq 1$.

Indeed, this is a remarkable result. It shows that the zeros of the Riemann zeta function of generalized Fermat function fields can be expressed by well known algebraic numbers which had appeared long ago already in classic algebraic number theory. The sums appearing on the right side of (6.14) are called **Jacobi sums**. They can be interpreted, after elementary computations, as factor systems of the **Gauss sums**

$$\tau(\chi) := - \sum_{a \in K^\times} \chi(a)e(a) \quad \text{with } e(a) := e^{\frac{2\pi i S(a)}{p}} \tag{6.15}$$

where $S : K \to \mathbb{F}_p$ denotes the trace function. Namely we have:

$$\omega_{\chi, \psi} = \frac{\tau(\chi)\tau(\psi)}{\tau(\chi\psi)} . \tag{6.16}$$

These factor systems had already appeared in Davenport's letter during his calculation for his estimate (see page 68). (Note the minus sign in the above definition (6.15) of τ. It had been inserted by Hasse in order to simplify the formulas.)

In view of the well known relation $|\tau(\chi)| = \sqrt{q}$ for Gauss sums, the RHp is an immediate consequence of the theorem. This gives a direct proof of the RHp for these fields, without recourse to Artin's criterion. Instead of Artin's criterion Hasse had used class field theory for global fields of characteristic p. He considered $F = K(x, y)$ as an abelian extension over the rational field $K(x^m)$, generated by two cyclic extensions according to (6.13):

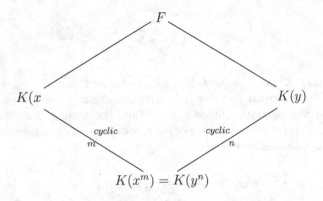

and he computed the corresponding L-series $L(s \mid \varphi)$ for ray class characters $\varphi \neq 1$ of $F \mid K(x^m)$. These L-series turn out to be polynomials in the variable $t = q^{-s}$, and they are the factors of F.K. Schmidt's polynomial $L(t)$ which occurs in (4.11) on page 47. This is seen by comparing (4.11) with the following relation known from class field theory

$$\zeta_F(s) = \zeta_{K(x^m)}(s) \cdot \prod_{\varphi \neq 1} L(s \mid \varphi) = \frac{\prod_{\varphi \neq 1} L(s \mid \varphi)}{(1 - q^{1-s})(1 - q^{-s})}$$

The ray class characters φ of $F \mid K(x^m)$ are products $\varphi = \chi\psi$ of the ray class characters χ of $K(x) \mid K(x^m)$ and ψ of $K(y) \mid K(y^n)$, and a detailed study of the situation, standard in class field theory, shows that these ray class characters can be identified with the characters χ, ψ of $K^\times$ appearing in the theorem above. It turns out that the corresponding L-series, if nontrivial, are *linear* polynomials (in the variable $t = q^{-s}$); their coefficients can be explicitly computed. This leads to formula (6.14).

REMARK The field degree $[F : K(x^m)] = mn$ is relatively prime to p. Therefore, in applying class field theory to the generalized Fermat fields, Hasse could have referred to F.K. Schmidt's paper [Sch31b] on class field theory in characteristic p which I have mentioned earlier (see page 42). However, F.K. Schmidt did not treat class fields of degree divisible by p. Hence Hasse decided to prepare an extra paper where he developed class field theory for global fields of characteristic p quite generally, without the degree restriction of F.K. Schmidt's paper. This paper included Artin's reciprocity law and the Local-Global Principle . Hasse obtained this by using Tsen's Theorem which states that over a function field with algebraically closed base field every central algebra splits [Tse33]. Tsen had recently obtained this result while working in Göttingen with Emmy Noether. Let me note in passing that the Local-Global Principle for algebras over function fields with finite base fields was also established in Witt's thesis [Wit34a], published in the same year 1934. Witt used the analytic theory of algebras which he had established in the function field case. He also proved the so-called "Existence Theorem" of class field theory in characteristic p.

Hasse's extra paper mentioned above appeared 1934 in Crelle's Journal [Has34e]. It is to be regarded not only as a preparation for the Davenport-Hasse paper [DH34] but it is also of high relevance for the general investigation of global fields in characteristic p. Hasse completed this paper *"within a few days"*, as he wrote in a letter to Davenport dated 15 May 1934. Actually, Hasse considered in [Has34e] cyclic extensions only. But the generalization to arbitrary abelian extensions is straightforward and was familiar in class field theory of that time.

6.3.3 Davenport-Hasse Fields

In the Davenport-Hasse paper [DH34] not only the generalized Fermat function fields defined by (6.9) are investigated, but also the function fields $F = K(x, y)$ defined by

$$y^p - y = x^m \tag{6.17}$$

where m is not divisible by the characteristic p . These fields are called the **Davenport-Hasse function fields**.

(Sometimes the terminology "Davenport-Hasse fields" is used for all types which were dealt with in the Davenport-Hasse paper [DH34], including the "generalized Fermat fields".)

In a certain sense the Davenport-Hasse fields are similar to the generalized Fermat fields, in as much as they are composite of two cyclic extensions:

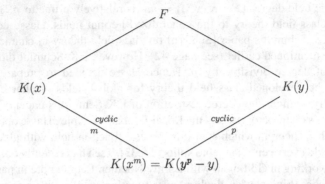

But now $y^p - y = x^m$ is an Artin-Schreier equation. Artin and Schreier had discovered in 1927 that equations of this type can be used to generate every cyclic extension of degree p in characteristic p [AS27b]. But prior to Hasse nothing was known about the *arithmetic* behavior of such extensions for global fields of characteristic p. Hasse entered completely unknown territory when he investigated such extensions. In particular their class field decomposition law had to be discovered, as well as the ramification behavior. Not even the genus was known.

Hasse found their genus to be

$$g = \frac{(p-1)(m-1)}{2}. \tag{6.18}$$

The following result similar to (6.14) was obtained:

Theorem *Let K be the finite field of characteristic p with q elements and $F = K(x, y)$ the Davenport-Hasse function field with the defining relation (6.17). It is assumed that m divides $q - 1$. Then the roots of F.K. Schmidt's L-polynomial are given by the generalized Gauss sums*

$$\tau_k(\chi) = -\sum_{a \in K} \chi(a) e^k(a). \tag{6.19}$$

where χ is a nontrivial character of $K^\times$ of order dividing m and $0 < k < p$. The exponential $e^k(a) = e(ka)$ is given by (6.15).

Again, class field theory from Hasse's paper [Has34e] was used in the proof, with respect to the abelian extension $F|K(x^m)$. A detailed study of the corresponding L-series led to the theorem. And again the RHp is an immediate consequence since $|\tau_k(\chi)| = |\tau(\chi)| = \sqrt{q}$.

It seems remarkable that for the Davenport-Hasse fields the Gauss sums themselves appear as the roots of the zeta function, whereas for the generalized Fermat fields the roots are *factor systems* of Gauss sums. In this respect the result for the Davenport-Hasse fields looks simpler. Accordingly, in the published paper [Has34c] the Davenport-Hasse fields are discussed first, and only thereafter the generalized Fermat fields are treated. I have changed the order of discussion since this reflects the historic line which is recognizable in the letters between Davenport and Hasse. After all, this work started with Davenport's letter of January 1932 on the congruence $ax^m + by^n \equiv 0 \mod c$. (See Sect. 6.3.1)

6.3.4 Stickelberger's Theorem

As said above, Hasse wished to give a full arithmetic characterization of the zeros of $L^*(t)$. With respect to the generalized Fermat fields and the Davenport-Hasse fields this comes down to the arithmetic characterization of Gauss sums, as seen in the foregoing sections.

Already in the year 1890 Stickelberger had solved this problem. He had determined the prime decomposition of Gauss sums, and in addition he proved certain congruence relations which allowed a complete characterization of Gauss sums by their arithmetic properties [Sti90].

Hilbert had included Stickelberger's Theorem in his *Zahlbericht* [Hil97] but only for prime numbers p. Hilbert did not consider the case of prime powers $q = p^r$. Of course Hasse had read Hilbert, in fact a whole generation of mathematicians had learned classical number theory from Hilbert's book. Accordingly Hasse knew Stickelberger's Theorem for prime numbers only. He was not aware of the fact that Stickelberger had also covered the case of prime powers.

So he sat down and produced, in cooperation with Davenport, the generalization to arbitrary prime powers $q = p^r$. The case of a prime p would be covering function fields over the prime field $\mathbb{F}_p$ only, whereas Davenport and Hasse dealt with function fields over an arbitrary finite field K. Both worked hard for this proof. But after finishing they were told by Mordell that Stickelberger had covered the case of prime powers $q = p^r$ as well, although Hilbert had not included this more general result in his book.

After some discussion in which Davenport advanced the point of accepting Stickelberger's old proof and Hasse proposed his new modernized proof, they decided to include both versions of the proof in their paper. Hasse's argument for including the new proof was:

> "...*ours* [the proof] *is more precise. Moreover the old proof and the whole matter seems to have slipped from the mind of our generation, presumably owing to Hilbert's inconceivable not giving it in his Zahlbericht.*"

For some time I have fostered the idea to present Stickelberger's Theorem in Hasse's version here—in order to prevent that the matter would slip again from the mind of today's generation. But then I abstained from this since it would lead us too far away from the main objective of this book, namely the history of the RHp. And Stickelberger's Theorem seems to be quite alive today and there is no danger of forgetting it. (See, e.g., [Lem00]). So I just refer to the Davenport-Hasse paper [DH34].

6.4   Exponential Sums

Let $f(x) \in K(x)$ be a rational function with coefficients in a finite field K with, say, q elements. The **exponential sum** for $f(x)$ is traditionally defined as

$$\Sigma_f := \sum_{a \in K \cup \infty} e(f(a)) \qquad (6.20)$$

where the exponential $e(\cdots)$ is as in (6.15) on page 71. Here a ranges over those elements in $K \cup \infty$ which are not poles of $f(x)$, i.e., for which $f(a) \in K$ is defined. For instance, if $f(x) \in K[x]$ is a polynomial then ∞ is the only pole and the condition is $a \in K$. Such exponential sums had been studied for several cases of $f(x)$, mostly when $q = p$ is a prime. For instance, if $f(x) = ax + bx^{-1}$ then Σ_f is called a "Kloosterman sum".

One of the classic problems had been to obtain a good estimate of Σ_f for $q \to \infty$. Already the early letters between Hasse and Davenport (and Mordell) show that they searched for good estimates but they were far from the conjectured best estimate of order $\sqrt{q}$.

During the preparation of Hasse's class field paper [Has34e] (which I have mentioned above already on page 73) he discovered a connection of the exponential sums (6.20) to the function field $F = K(x, y)$ defined by

$$y^p - y = f(x) \qquad (6.21)$$

when F is considered as a cyclic Artin-Schreier extension of the rational function field $K(x)$. In particular the class field structure of this extension is of importance. Prior to Hasse nothing was known about it.

Thanks to the Hasse-Davenport correspondence which has been preserved, it is possible to follow step by step how Hasse conquered this unknown territory. On 24 July 1933 Hasse wrote:

> "For $y^p - y = f_3(x)$ [polynomial of degree 3] the genus is really $p - 1$. Further I can explicitly give the characters for any $y^p - y = f(x)$ (polynomial) …I have reason to suppose that my method for determining the genus applies to all cases where $f(x)$ is a rational function of "degree" less than p."

Already one day later Hasse reported:

"*I have got much more general results on*

$$y^p - y = f(x)$$

than I first thought. As a matter of fact, I have determined the genus, and with it the number of zeros of the corresponding L–function for every $f(x)$ (integer or fractional). If $f(x) = f_1(x) + \cdots + f_r(x)$ is the decomposition of $f(x)$ into partial fractions, and if all terms out of these $f_i(x)$ which are pure p^{th} powers are removed by an easy transformation of y, and if $n_1, \ldots, n_r$ are the degrees [of the pole divisors] of those $f_1(x), \ldots, f_r(x)$, then

$$g = \frac{(p-1)(n+r-2)}{2} \quad \text{where} \quad n = n_1 + \cdots + n_r \qquad (6.22)$$

and therefore the L-functions have $n + r - 2$ zeros."

Hasse means the L-functions for ray class characters of $F|K(x)$. These are polynomials (in the variable q^{-s}). There are $p - 1$ such L-functions and their product is F.K. Schmidt's $L(t)$. In his class field paper [Has34e] Hasse computes the relevant coefficients of these L-functions and shows that the sum of their inverse roots coincides with the exponential sum above. More precisely, since there are $p-1$ such L-functions there arise $p - 1$ exponential sums, each belonging to a power e^k of the exponential function, $0 < k < p$:

$$\Sigma_f^{(k)} := \sum_{a \in K \cup \infty} e^k(f(a)). \qquad (6.23)$$

But since these are all conjugate it suffices to consider Σ_f (for $k = 1$). Hasse says in [Has34e]:

"*...the problem of estimating the sums Σ_f is ...definitely solved once the RHp for $F|K$ is solved.*"

Indeed, since the L-functions have $n + r - 2$ roots and since these are also roots of F.K. Schmidt's $L(t)$, the RHp implies

$$|\Sigma_f| \leq (n + r - 2)\sqrt{q}. \qquad (6.24)$$

In particular, for Kloosterman sums it follows that $|\Sigma_f| \leq 2\sqrt{q}$. This order of magnitude $\mathcal{O}(q^{\frac{1}{2}})$ (for $q \to \infty$) is much better than the estimates which were known at the time, e.g., $\mathcal{O}(q^{1-\frac{1}{n}})$.

Today this can be found in almost every textbook on algebraic function fields (see, e.g., [Sti09]). But when Hasse prepared his paper [Has34e] he had to discover this (and more) himself.

To be sure, the RHp had not yet been generally proved at the time when Hasse wrote the lines above. He could solve the RHp for the case $f(x) = x^m$ only, see Sect. 6.3.3. But in my opinion, already the discovery of the connection between exponential sums and the L-functions of an Artin-Schreier extension is to be regarded as an outstanding advance in the historical development. Note that most of this work on Davenport-Hasse fields was done parallel to the work on elliptic function fields which I shall report on in the next chapter.

Summary

In 1931 Hasse met the young student Davenport who had been recommended to him by Mordell. A friendship developed between the two, and Hasse became interested in Davenport's work on estimating the number of solutions of diophantine congruences. Some months later Mordell extended and generalized Davenport's results. Although Hasse appreciated the high value and the ingenuity of Mordell's and Davenport's methods he voiced his opinion that the structural methods of "modern algebra" would lead to better estimates. Davenport challenged him to solve the problem with abstract structural methods. Hasse accepted this challenge and started to work on the problem, looking for the adequate algebraic structure connected with it.

In November of 1932 Hasse delivered a colloquium lecture in Hamburg and met Artin there. Artin told him about his results on the RHp which he had written in the year 1921 in a letter to Herglotz but never published. In this way Hasse learned that the Davenport-Mordell problem for diophantine congruences, if extended to arbitrary finite base fields instead of the prime field modulo p, is in fact equivalent to the RHp. Thus Hasse saw the structure behind the Davenport-Mordell problem, namely the theory of algebraic function fields in characteristic p and their zeta functions.

In early January of 1932 Davenport had written a letter to Hasse containing an estimate for the number of solutions of the generalized Fermat congruence $ax^m + by^n \equiv c \bmod p$ where p is a prime number. It turned out that this number is approximately p with an error term of order of magnitude $\sqrt{p}$. This constituted the best possible result which was to be expected. Later Hasse observed that the result can be generalized over an arbitrary finite field $\mathbb{F}_q$ with the error term estimated by $\sqrt{q}$ for $q \to \infty$. After his visit to Artin in Hamburg in November of 1932 Hasse became aware that this observation yields a proof of the RHp for the corresponding function field in characteristic p.

Moreover, the zeros of the corresponding zeta function (if considered as function of the variable $t = q^{-s}$) can be described by factor systems of Gauss sums. This led to a joint paper of Davenport and Hasse which appeared in 1934. In the same paper the authors dealt with function fields defined by an Artin-Schreier equation of the form $y^p - y = x^m$. The zeros of the zeta function of such a field are described by Gauss sums. Today those fields are called Davenport-Hasse fields.

In their proofs the authors used class field theory for global fields of characteristic p, including L-functions for the corresponding ray class characters. At that time this theory had not yet been fully developed in the literature, hence Hasse published another paper containing full proofs of the relevant class field theory for function fields, including Artin's reciprocity law.

Chapter 7
Elliptic Function Fields

7.1 The Breakthrough

We all know that a good way to study a mathematical subject is to give a lecture course about it. The necessity to arrange the theory in a systematic way and to explain to the audience the various connections between the different results, often leads to new insights and, in consequence, to new results.

In this spirit Hasse had lectured in his seminar in February of 1933 on function fields in characteristic p, in particular on elliptic fields, i.e., when the genus $g = 1$. In the same month he succeeded with the proof of RHp in the elliptic case. This result got a lot of attention in the mathematical community.

Davenport wrote to Hasse on 21 February 1933:

"... I am much excited as to whether your new idea for $y^2 \equiv f_3(x)$ comes off. The result for $f_3(x, y) \equiv 0$ follows from it without further work."

Here, f_3 denotes a polynomial of degree 3. (It is amusing that Davenport always writes $y^2 = f_3(x)$ when he is talking about elliptic curves or function fields, whereas Mordell prefers $y^2 = f_4(x)$. Of course, every function field of genus 1 and characteristic $\neq 2$ can be generated by both of these normal forms if it contains a prime of degree 1, which is the case if the base field is finite. Hasse, in his correspondence with Davenport and with Mordell, used $f_3(x)$ or $f_4(x)$ respectively, depending on his addressee.)

From Davenport's wording it is not certain that Hasse at that time had already completed his proof. Davenport's words could also be interpreted such that Hasse had given him a rough outline of his ideas without having them worked out already. This interpretation seems not to be unrealistic because in the next letter of Davenport we read:

"... I am waiting with great eagerness to hear what the final result of your work will be. It will be a marvellous achievement, and should lead to

© Springer Nature Switzerland AG 2018
P. Roquette, *The Riemann Hypothesis in Characteristic* p
in Historical Perspective, Lecture Notes in Mathematics 2222,
https://doi.org/10.1007/978-3-319-99067-5_7

the solution of other problems, i.e., Kloosterman sums, which are closely connected to $y^2 \equiv f_3(x)$. I re-read your letter in which you explained your method the other day... I hope in a few days I shall be able to congratulate you on a final solution of the problem ..."

In a letter dated 6 March 1933 Hasse reported his new result to Mordell. That letter is preserved. It leaves no doubt that at this date Hasse was in possession of the proof:

"Dear Prof. Mordell, I succeeded recently in proving that the number of solutions of $y^2 \equiv f_4(x)$ mod p is $p + term$ which is less than or equal to $2\sqrt{p}$. Moreover, the same holds for any Galois–field instead of rational Galois–field mod. p, that is, the analogue to Riemann's hypothesis is true for the corresponding Artin Zetafunction ..."

Mordell's reply to Hasse's letter is dated 9 March 1933:

"... I was exceedingly interested in your mathematical news and was very glad to hear that you had completely knocked down the bottom out of $y^2 \equiv f_4(x)$ mod p. It is a wonderful achievement and I shall look forward with the greatest interest to seeing your paper in print. I hope you will make it as easy as possible for the reader to understand, without reference to all the theorems on Klassenkörpertheorie etc. For as this is the first case of any exact result for zeros of Zetafunctions on $\Re(s) = \frac{1}{2}$, the paper is sure to attract an enormous amount of attention ...

What a tremendous vindication (for those who need it and have not appreciated the K.k.theory) that the proof should depend upon such a comparatively high brow theory. I feel rather relieved to think I did not spend too much time on further results of this kind with my method, and very pleased that my old paper should have supplied even an amount ε of usefulness ..."

Mordell refers to his "old" paper [Mor22] which contains Mordell's part of what today is called "Mordell-Weil Theorem".

The sentences about class field theory ("*K.k.theory*") reflects that Mordell was no friend of what he called "high brow theory". In this case, however, he seems to have accepted it. To him Hasse's result on the RHp carries sufficient "vindication". In his first proof Hasse had used class field theory over imaginary quadratic fields in the framework of classic complex multiplication (see Sect. 7.3).

Perhaps it was Hasse's use of class field theory in his first proof which later, in October of 1933, induced Mordell to ask Hasse about a possible English translation of Hasse's 1932 Marburg Lecture Notes on class field theory [Has33b]. Those notes contained the foundations of general class field theory according to the state of the art at the time. They had been mimeographed and distributed among interested mathematicians. Perhaps Mordell wanted his British colleagues and students to learn class field theory in order that they would be able to follow Hasse's proof of the RHp. (Mordell himself did not need a translation since he had a fairly good knowledge of German.) Hasse consented to the translation and recommended

Reinhold Baer as someone who would well be able to do it.. Moreover, Baer was to edit a follow up of Hasse's Lecture Notes, covering Hasse's new theory of norm residues; this part had not yet been included in the Notes. In the end Baer and Mahler were designated for the translation.

SIDE REMARK Reinhold Baer had to leave Germany due to the antisemitic policy of the German Nazi government, and he had gone to England after Hasse had recommended him to Mordell. Hasse held a friendship with Baer from their time as colleagues in Halle—a friendship which lasted a lifetime. Hasse had hoped that Baer through the translation job would get some additional financial support.

But in spring of 1934 the plan of translation of Hasse's Marburg Lecture Notes was abandoned, for reasons not known to me. (Many years later Hasse's Marburg lectures were reprinted in book form, see [Has67].) One reason may have been that Hasse had changed his proof of the RHp for elliptic fields. The new proof did not use class field theory. What kind of change was this, and why did Hasse perform this change? Let us see:

7.2 The Problem

Let F be an elliptic function field with finite base field K, and let $q = p^r$ be the order of K. Since the genus of F is $g = 1$, F.K. Schmidt's L-polynomial of F is quadratic (see formula (4.13) on page 47):

$$L^*(t) = t^2 + (N - q - 1)t + q = (t - \omega_1)(t - \omega_2), \qquad (7.1)$$

and the RHp for F implies that $|N-q-1| \leq 2\sqrt{q}$. This means that the discriminant satisfies

$$\Delta := (N - q - 1)^2 - 4q \leq 0, \qquad (7.2)$$

i.e., that the roots ω_1, ω_2 are contained in an imaginary quadratic field. Conversely, assume that (7.2) holds. If $\Delta < 0$ then $\mathbb{Q}(\omega_1) = \mathbb{Q}(\omega_2)$ is an imaginary quadratic field and ω_1 is conjugate to ω_2. Since $\omega_1\omega_2 = q$ the RHp follows: $|\omega_1| = |\omega_2| = \sqrt{q}$. If $\Delta = 0$ then $\omega_1 = \omega_2 = \pm\sqrt{q}$ and hence again $|\omega_1| = |\omega_2| = \sqrt{q}$. We see that the RHp requires to solve the following

Problem *Given an elliptic function field $F|K$ with finite base field K containing q elements, let N denote the number of prime divisors of degree 1 in $F|K$. Find an imaginary quadratic number field $\Omega|\mathbb{Q}$ and an element $\pi \in \Omega$ such that*

$$\mathcal{N}(\pi) = q \quad \text{and} \quad \mathcal{S}(\pi) = -(N - q - 1) \qquad (7.3)$$

where $\mathcal{N}$, $\mathcal{S}$ denote norm and trace for $\Omega|\mathbb{Q}$.

I am writing the norm operator from the quadratic field Ω with a calligraphic $\mathcal{N}$, in order to distinguish it from the number N of primes of degree 1 of $F|K$. Accordingly the trace is written as calligraphic $\mathcal{S}$ ("Spur" in German).

If the above problem is solved the roots ω, ω' of $L(t)$ can be identified with π and its conjugate $\bar{\pi}$.

Note that $\mathcal{N}(\pi-1) = \mathcal{N}(\pi)-\mathcal{S}(\pi)+1$, hence the relations (7.3) are equivalent to

$$\mathcal{N}(\pi) = q \quad \text{and} \quad \mathcal{N}(\pi - 1) = N . \tag{7.4}$$

Today this problem is readily solved as follows:

Assume that $F = K(\Gamma)$ be the function field of an elliptic curve Γ defined over K. In view of the well known addition formulas we may regard Γ as an abelian variety of dimension 1. The K-rational points of Γ form a finite subgroup $\Gamma(K)$, and its order equals the number N of prime divisors of degree 1 of F. Consider the endomorphism ring $M(\Gamma)$ in the sense of algebraic geometry. It is known that this endomorphism ring is (isomorphic to) a subring of an imaginary quadratic field Ω (up to a few exceptional cases). Every endomorphism $\mu \in M(\Gamma)$ has a degree, and this degree equals the norm $\mathcal{N}(\mu)$ from Ω to $\mathbb{Q}$. If μ is separable then the degree equals the order of the kernel of μ.

Now, the q-*Frobenius operator* $\pi \in M(\Gamma)$ is defined as raising the coordinates of any point of Γ to their q-th power. Applied to a generic point, we see that its degree is $[F : F^q] = q$, which gives the first relation in (7.4). On the other hand, by definition π fixes precisely the K-rational points; hence $\Gamma(K)$ is the kernel of $\pi - 1$. This gives the second relation in (7.4) since $\pi - 1$ is separable. In other words:

F. K. Schmidt's polynomial $L^*(t)$ is the characteristic polynomial of the Frobenius operator within the endomorphism ring $M(\Gamma)$, and the latter is a subring of an imaginary quadratic field Ω.

This sounds "*Quite trivial*" if I am allowed to use the same words which Davenport had exclaimed at the end of his proof in his letter of January 1932 (see page 69).

The "few exceptional cases" mentioned above are those where Γ is "supersingular" in the sense of Deuring, i.e., where $M(\Gamma)$ is non-commutative. See Sect. 8.4.1. Those elliptic curves had been discovered by Hasse. They can be characterized by the vanishing of a certain number A, the so-called Hasse invariant of Γ. (I shall explain the definition of A in Sect. 8.1.) In these cases $M(\Gamma)$ is a maximal order in the quaternion algebra which is ramified at p and ∞ only. Every maximal commutative subfield of this quaternion algebra is imaginary quadratic. Since the Frobenius endomorphism π is contained in such a maximal subfield, the same argument as above can be applied in this case. (Here, $\mathcal{N}$ equals the reduced norm from the quaternion algebra.)

But these arguments were not yet available when Hasse started his work in 1933. The standard notions and results about elliptic curves in characteristic p and their endomorphism rings which I have used here, were unknown at the time. In fact, Hasse had to create these notions and to prove their relevant properties during the next years. What I have sketched above constitutes Hasse's second proof. But what was his first proof and why did he refrain from publishing that?

Dear reader, if you just want to know how Hasse's second proof works in the elliptic case then you may be satisfied with what I have said above. You may skip the next sections where I shall talk about the gradual evolution of the ideas which led Hasse finally to what was named "complex multiplication in characteristic p", and what became the base of the arguments employed above. We have here one of the rare situations where this evolution is well documented by letters and notes.

7.3 Hasse's First Proof: Complex Multiplication

The theory of complex multiplication has a long history back into the nineteenth century. In the 1920s Hasse had published two important papers on the foundations of complex multiplication in connection with Takagi's class field theory [Has26b, Has31]. So he was familiar with the close relation between elliptic curves in characteristic 0 and the arithmetics of imaginary quadratic fields, since this is the subject of complex multiplication. Therefore it is understandable that when Hasse first encountered the problem (7.4) he remembered his own past work and wondered whether there may be a connection of complex multiplication to the problem (7.3) of RHp.

The classic theory of complex multiplication is concerned with elliptic curves in characteristic 0. Accordingly Hasse's first idea was to lift the problem from characteristic p to characteristic 0. The original elliptic curve Γ in characteristic p should be represented, if possible, as a good reduction of an elliptic curve Γ^* in characteristic 0. (You may have noticed that here I am switching from to the language of function fields to the language of algebraic geometry.) At that time, however, the general notion of "good reduction" had not yet been defined, and Hasse had to perform certain computations directly which today are standard as immediate consequences of the theory of good reduction. The general theory was introduced later by Deuring [Deu42] while generalizing and systematizing Hasse's arguments. See Sect. 8.4.4.

The lifted curve Γ^* should have suitable properties, suitable for the solution of problem (7.4). In order to describe Hasse's lifting process it may be helpful first to remember the relevant facts from classic complex multiplication.

Let Γ be an elliptic curve defined over any field K. Consider its modular invariant, usually denoted by $j = j(\Gamma)$, which is a non-zero element in the base field K. If the characteristic is $\neq 2, 3$ and if Γ has a K-rational point then the curve can be defined by an equation in Weierstraß normal form:

$$y^2 = 4x^3 - g_2 x - g_3 \qquad \text{with } g_2, g_3 \in K \,, \tag{7.5}$$

and then j is defined as

$$j = 12^3 \frac{g_2^3}{\Delta} \qquad \text{with } \Delta = g_2^3 - 27 g_3^2 \,. \tag{7.6}$$

This is a birational invariant of the curve (over the algebraic closure of the base field K).

If the base field is the complex number field $\mathbb{C}$ then the elliptic curve Γ can be analyrically uniformized by the additive factor group $\mathbb{C}/\mathfrak{w}$ where $\mathfrak{w}$ denotes the lattice of periods appearing when the unique differential of the first kind is integrated. The period module $\mathfrak{w}$ is uniquely determined by Γ up to a scalar factor. In this situation g_2, g_3 and hence j can be given by analytical formulas as follows:

$$g_2 = g_2(\mathfrak{w}) = 60 \sum_{0 \neq w \in \mathfrak{w}} \frac{1}{w^4} \,, \qquad g_3 = g_3(\mathfrak{w}) = 140 \sum_{0 \neq w \in \mathfrak{w}} \frac{1}{w^6} \,. \tag{7.7}$$

The defining relation (7.5) of the curve is then satisfied by the Weierstraß $\wp$-function and its derivative:

$$x = \wp(z|\,\mathfrak{w}) = \frac{1}{z^2} + \sum_{0 \neq w \in \mathfrak{w}} \left(\frac{1}{(z-w)^2} - \frac{1}{w^2} \right) \tag{7.8}$$

$$y = \wp'(z|\,\mathfrak{w})$$

where z is a complex variable and differentiation is with respect to z. The function $\wp(z|\,\mathfrak{w})$ (as a function of z) is meromorphic with a pole of order 2 at $z = 0$, and it is periodic with the period module $\mathfrak{w}$. The map

$$z \mapsto (\wp(z|\,\mathfrak{w}), \wp'(z|\,\mathfrak{w})) \tag{7.9}$$

establishes a bijection between the factor group $\mathbb{C}/\mathfrak{w}$ and $\Gamma(\mathbb{C})$. Due to this bijection $\Gamma(\mathbb{C})$ is equipped with the structure of an additive group with the point at infinity as the zero element. Let us write $\Gamma^{\mathfrak{w}}$ for the elliptic curve analytically uniformized by $\mathbb{C}/\mathfrak{w}$ in this way.

A *multiplier* of $\mathfrak{w}$ is defined to be a complex number μ such that $\mu\mathfrak{w} \subset \mathfrak{w}$. This defines an endomorphism, likewise denoted by μ, of the factor group $\mathbb{C}/\mathfrak{w}$ and hence of $\Gamma^{\mathfrak{w}}(\mathbb{C})$. The ring $\mathsf{M} = \mathsf{M}(\Gamma^{\mathfrak{w}})$ of all those multipliers contains $\mathbb{Z}$. If there are more multipliers in M then these are not real numbers. In this case the lattice $\mathfrak{w}$ (or the elliptic curve $\Gamma^{\mathfrak{w}}$) is said to admit *complex multiplication*. From this the whole theory derives its name.

Assume that $\Gamma^\mathfrak{w}$ admits complex multiplication. Then the multiplier ring M, acting on the period module $\mathfrak{w}$ which is 2-dimensional, is seen to be a so-called "order" of some imaginary quadratic number field Ω, i.e. a subring of the integers which generates the field Ω. (Today the word "order" in this connection is not in general use any more.) In this case the lattice $\mathfrak{w}$ is proportional to some lattice in Ω and one may assume $\mathfrak{w} \subset \Omega$. Then $\mathfrak{w}$ appears as an ideal (integer or fractional) of M. In general M is not integrally closed, i.e., it may admit a nontrivial conductor. This is a natural number m, and M consists precisely of all algebraic integers in Ω which are congruent to some rational integer modulo m. (Hasse says "index" instead of "conductor", following the classic terminology in the theory of complex multiplication.) After adjusting $\mathfrak{w}$ by a suitable proportionality factor one may assume that $\mathfrak{w}$ is relatively prime to the conductor. All ideals of M which are relatively prime to the conductor form a multiplicative group. Modulo principal ideals we obtain the *ideal class group* of M; this is finite.

In this situation, the invariant $j(\mathfrak{w})$ of $\Gamma^\mathfrak{w}$ turns out to be an algebraic number generating an abelian extension $\mathfrak{K}$ of Ω. In fact, $\mathfrak{K}$ is the so-called *ring class field* of M in the sense of class field theory, and $j(\mathfrak{w})$ depends on the ideal class of $\mathfrak{w}$ only. If $\mathfrak{w}$ ranges over the ideal class group of M then the corresponding invariants $j(\mathfrak{w})$ are different, and they constitute a complete set of conjugates over Ω. The Galois group of $\mathfrak{K}|\Omega$ is isomorphic to the ideal class group of M via Artin's reciprocity law. The curve $\Gamma^\mathfrak{w}$ can be defined over $\mathfrak{K}$ (up to birational equivalence).

This being said, we can now formulate Hasse's

Lifting Theorem *Let K be a finite field of characteristic p with q elements, and Γ an elliptic curve defined over K with invariant $j \in K$. Assume j to be of degree r, so that $\mathbb{F}_p(j) = \mathbb{F}_{p^r} \subset K$. There exists an imaginary quadratic field $\Omega \subset \mathbb{C}$ and a lattice $\mathfrak{w}$ in Ω such that:*

1. *p does not divide the conductor of the multiplier ring $\mathsf{M}(\Gamma^\mathfrak{w})$.*
2. *p admits a prime divisor $\mathfrak{P}$ of degree r in the ring class field $\mathfrak{K} = \Omega(j(\mathfrak{w}))$ such that, after suitable identification of the residue field of $\mathfrak{K}$ mod $\mathfrak{P}$ with $\mathbb{F}_p(j)$ we have $j(\mathfrak{w}) \equiv j$ mod $\mathfrak{P}$.*
3. *q splits in Ω into two factors: $q = \pi\pi'$.*

In general the lattice $\mathfrak{w}$ is uniquely determined by Γ up to a scalar factor from Ω. Exceptional cases arise only in few cases, namely when the so-called Hasse invariant A of Γ vanishes; in such case there are precisely two non-equivalent lattices $\mathfrak{w}, \mathfrak{w}'$ satisfying the conditions 1–3.

I shall explain later the definition and relevance of the Hasse invariant which is usually denoted by A. See Chap. 8.

Actually, at the time when Hasse formulated the Lifting Theorem he had not yet discovered the invariant A. He proved the Lifting Theorem only under the additional condition that $p \neq 2, 3$ and, moreover, the degree r of j should be odd. He believed this last condition to be unnecessary and, indeed, this has been verified later by Shiratani [Shi67]. Today it is known that the Hasse invariant A may vanish only if the degree $r = 1$ or $r = 2$. The obstruction which Hasse encountered in his proof when r is even is related to the possible vanishing of A for $r = 2$. See Chap. 8 below.

In view of Hasse's Lifting Theorem above we can understand what Hasse had meant when he wrote to Mordell about "uniformizing a congruence", as he did in his letter of 6 March 1933:

"It is a curious fact that the leading idea of my proof may be considered as the fruit from our reading Siegel's great paper last year, or rather of my learning your method in the elliptic case.   For, as there the equation $y^2 = f_4(x)$ *is treated by uniformizing it through elliptic functions, so I now treat the congruence* $y^2 \equiv f_4(x)$ *mod.p by uniformizing it the same way..."*

Here, Hasse refers to Mordell's visit 1 year earlier, during the Easter vacations 1932, when the Mordells had stayed with the Hasses in Marburg. This visit had happened just at the time when Hasse had to write a *Jahrbuch review* of Siegel's paper [Sie29], and he used Mordell's presence to go over Siegel's paper jointly page by page. (This review has the unusual size of more than 5 pages. It appeared in volume 56 of the *Jahrbuch für die Fortschritte der Mathematik*. See footnote 4 on page 4.)

It seems that on this occasion Mordell had explained to Hasse his use of the uniformization of elliptic curves in his old paper [Mor22] which contained Mordell's part of the "Mordell-Weil Theorem". When Hasse now speaks of "uniformization of the congruence" then this can be interpreted such that the curve Γ in characteristic p should be lifted to the curve $\Gamma^{\mathfrak{w}}$ with invariant $j(\mathfrak{w})$, and then the said "uniformization" of Γ is $\mathbb{C}/\mathfrak{w}$, or rather its torsion subgroup. (By the way, Mordell once told me that he did not like the terminology "Mordell-Weil Theorem." In his opinion he and Weil had proved two different theorems which should be referred to as "Mordell's Theorem" and "Weil's Theorem" respectively.)

It is not my aim here to go into the details of proof of Hasse's Lifting Theorem. I have mentioned the theorem in order to point out the general direction of the ideas of Hasse's first proof. Hasse tried to connect the RHp with the classical theory of complex multiplication which he was familiar with.

Observe that the Lifting Theorem alone does not yet give a proof of problem (7.4). While property 3. of the Lifting Theorem yields an element π with $\mathcal{N}(\pi) = q$, the second condition $\mathcal{N}(\pi - 1) = N$ of (7.4) remains to be verified. To this end, Hasse had to extend the ring class field $\mathfrak{K}$ to the so-called $(\pi - 1)$-division field $\mathfrak{K}'$ which is generated over $\mathfrak{K}$ by the division values $\wp(\alpha), \wp'(\alpha)$ where $\wp$ denotes the Weierstrass $\wp$-function belonging to $\mathfrak{w}$ and α ranges over the complex numbers modulo $\mathfrak{w}$ which are annihilated by $\pi - 1$. i.e., $\pi\alpha \equiv \alpha$ mod $\mathfrak{w}$. The points of Γ have to be lifted to points of $\Gamma^{\mathfrak{w}}$ and those are represented by the values $\wp(\alpha)$ and $\wp'(\alpha)$ as its coordinates. By definition π acts on the points of $\Gamma^{\mathfrak{w}}$, hence on the values $\wp(\alpha)$ and $\wp'(\alpha)$ as their coordinates. In this way it induces an isomorphism φ_π in $\mathfrak{K}'|\mathfrak{K}$. Since $\pi\alpha \equiv \alpha$ mod $\mathfrak{w}$ it is seen that this isomorphism is the identity on $\mathfrak{K}'$. On the other hand, $\mathfrak{K}'$ is an abelian extension of $\mathfrak{K}$, hence can be viewed as class field. Artin's reciprocity law maps the prime ideal $\mathfrak{P}$ appearing in the lifting theorem onto a certain automorphism $\varphi_{\mathfrak{P}}$ of $\mathfrak{K}'|\mathfrak{K}$. In fact, $\varphi_{\mathfrak{P}}$ is the

Frobenius automorphism belonging to $\mathfrak{P}$, which maps every element onto its q-th power modulo $\mathfrak{P}$. The theory of complex multiplication asserts that $\varphi_{\mathfrak{P}} = \varphi_{\pi}$, hence $\varphi_{\mathfrak{P}}$ induces the identity in $\mathfrak{K}'$. It follows that $\wp(\alpha)^q \equiv \wp(\alpha) \bmod \mathfrak{P}$ for every $\alpha \in \frac{1}{\pi - 1}\mathfrak{w}$. Similarly for $\wp'(\alpha)$ modulo $\mathfrak{P}$. Reducing modulo $\mathfrak{P}$ it follows that these points are mapped onto those points (x, y) of Γ for which $x^q = x$, $y^q = y$, i.e., the K-rational points of Γ. This then leads to the second relation of (7.4).

Actually, Hasse worked here not directly with the Weierstrass function $\wp(z)$ but with some modification of it, called "Weber's function" $\tau(z)$ which is more adapted to arithmetical investigations. But here I do not wish to go into the details. My aim is to point out the main ideas for Hasse's first proof.

The rising of these ideas is well documented. In his letter to Mordell of 6 March 1933 Hasse gave a brief sketch of his main ideas. In his *Nachlass* there are lecture notes for a seminar lecture in Marburg in May of 1933, obviously meant as a continuation of Hasse's seminar talk in February which I have mentioned above already (page 81). Another outline of his first proof is published in the "*Göttinger Nachrichten*" of 1933 [Has33a]. In the *Nachlass* of Hasse there was found a manuscript with a complete proof, ready for publication but never submitted.[1] Hasse had given colloquium talks about it in several places, including the annual meeting of the DMV in September of 1933 at Würzburg. His report about this talk in Würzburg appeared in the *Jahresbericht der DMV* [Has34c].

But while writing down this report Hasse became more and more convinced that lifting of the problem to characteristic 0 is not really necessary, and most of the arguments would work in characteristic p already if suitably formulated. On 5 November 1933 he wrote to Davenport:

> "*I ought to make my treatment of the elliptic case by means of elliptic functions ready for print. But I cannot get myself to working at it. I rather should like to avoid this publication at all by giving a pure algebraic proof. I have made some definite progress in this direction in the last weeks.*"

When Hasse submitted the text of his Würzburg report for printing he had already decided to rewrite the proof as indicated in his letter to Davenport. Accordingly, he added the following announcement:

> "*In the meantime I succeeded to give this proof for all elliptic cases in characteristic different from 2 in a purely algebraic way. I am developing a purely algebraic theory of division fields and of complex multiplication of elliptic function fields with algebraically closed base field. In my opinion this new proof seems to be more adapted to the situation. Hence I will refrain from the presentation of the first analytic proof. The publication of the second purely algebraic proof will appear in the near future as a sketch in the Proceedings of the Mathematical Seminar Hamburg, and in detail in Crelle's Journal.*"

---

[1]The manuscript has 94 pages. I am indebted to Reinhard Schertz for providing me with a copy.

7.4 The Second Proof

In the announcement above Hasse refers to his forthcoming paper [Has34a] in the
Hamburger Abhandlungen which contains the Lecture Notes of his colloquium
lectures which he had delivered in Hamburg in January 1934. There he had
introduced the main ideas of his new proof. He had been invited by Artin who
apparently had heard about the new proof. Artin had written:

> *"Dear Mr. Hasse! Wouldn't you want to come over here this semester and
> give a talk? You could talk abut whatever you want to. Perhaps the beautiful
> results on the Riemann conjecture for function fields ? After all, this is the most
> beautiful result that has been obtained in the last few decades. My students
> would be highly interested. It would be nice if you could take off a whole week
> ..."*

Hasse accepted this invitation and went to Hamburg for a week. There he
delivered three lectures. The audience which Artin had mentioned in his letter were
probably the members of his seminar. But the lectures were announced as public,
and certainly there were other people too attending Hasse's lectures. Very likely the
following persons were present:

- *Max Zorn.* He had obtained his Ph.D. in 1930 with Artin who considered him as
  one of his most brilliant students. Thereafter he got a position at the University of
  Halle as an assistant to H. Brandt, the successor of Hasse there. In 1932 he quitted
  this position and moved to Hamburg again. In the year 1933 there appeared his
  paper in the *Hamburger Abhandlungen* which showed that the thesis [Hey29] of
  Käte Hey (the first Ph.D. student of Artin) could be interpreted so as to yield a
  proof of the Local-Global-Principle for algebras. This paper [Zor33] had received
  great interest among the people working in class field theory, including Emmy
  Noether and Hasse. In 1934 Zorn emigrated to the USA. (Today his name is
  known through Zorn's Lemma).
- *Wei-Liang Chow* who had studied in Göttingen with Emmy Noether but now
  planned to change to Leipzig in order to work with van der Waerden. He resided
  mainly in Hamburg (where he had found his later wife Margot) and, as reported
  by Chern [Cea96], kept close contact to Artin. (Chern himself was probably not
  present at Hasse's lectures; according to his own testimony he came to Hamburg
  in October of 1934.)
- *Hans Petersson.* He had been a Ph.D. student of Hecke. Now he held a position
  as *"Privatdozent"* at Hamburg University. He had lectured on class field theory a
  year before. He was a referee for the Chevalley-Weil paper [CW34] which also
  appeared in the *Hamburger Abhandlungen* of the current year 1934. That paper
  gives an important contribution to the algebraic theory of function fields; it is
  evident that this topic is closely connected to the program which Hasse presented
  in his Hamburg lectures.
- *Harald Nehrkorn*, a Ph.D. student of Artin in 1933 who in his thesis [Neh33]
  provided algebraic proofs of Artin's class number relations. In the 1935 volume

of the *Hamburger Abhandlungen* he published a paper jointly with Chevalley on class field theory [CN35], taking a big step towards a purely algebraic foundation of Artin's reciprocity law.

- *Heinz Söhngen*, another Ph.D. student of Artin, in 1934. His thesis [Söh35] is about complex multiplication, expanding a former paper of Hasse. (Later he went to applied mathematics.)
- *Walter Landherr*, also a Ph.D. student of Artin in 1934. His thesis [Lan35] dealt with simple Lie rings over p-adic fields.
- *Hans Zassenhaus*, yet another Ph.D. student of Artin in 1934, working in group theory. Even before he received his Ph.D. degree, his name became known through his explicit proof [Zas34] of the Jordan-Hölder-Schreier Theorem in group theory (the Butterfly Lemma). From the beginning of his mathematical career he was very active in group theory; we have counted 7 important published papers in the years 1934–1938. For a long time his text book [Zas37] was considered as "the" classical introduction to group theory.
- *Erich Kähler* who had obtained his Ph.D. in Leipzig but now worked in Hamburg where had got his *Habilitation* with Blaschke in 1930, the same year as Hans Petersson. His interest was mainly with differential geometry.
- *Erich Hecke* and *Wilhelm Blaschke*, Artin's colleagues in Hamburg.

This list of names, certainly not complete, shows that Hasse met a highly competent and interested audience in Hamburg. As Hasse had announced, the notes for these lectures appeared in the *Hamburger Abhandlungen* [Has34a]. (That issue contains the photographic portrait of Hasse, which is shown at the beginning of Chap 5. I do not know who had made it. Perhaps Natascha Artin? The photo is mentioned in a letter of 26 February 1934 from Blaschke to Hasse.)

Another preliminary report, somewhat amended, appeared one year later in the *Göttinger Nachrichten* [Has35]. In particular Hasse could now include the case of characteristic 2. The final proof was submitted for publication in November 1935 and appeared 1936 in Crelle's Journal in three parts [Has36c]. Due to these documents, and from Hasse's letters to Davenport, we can observe how the ideas and methods for the second proof took shape gradually in the course of time.

Hasse in his first proof, while working with the reduction modulo $\mathfrak{P}$ of the $(\pi - 1)$-division field (see page 88), observed that the quadratic integer $\pi \in M$ after reduction acts on Γ such that $\pi - 1$ kills precisely the K-rational points of Γ. This induced him to build his new proof as follows:

Step 1. Algebraic construction of the endomorphism ring of the Jacobian of an elliptic function field over an arbitrary base field, in particular over fields of characteristic $p > 0$. (Hasse says "ring of multipliers" of the elliptic curve as it was usual in the classical case.) Creation and definition of the necessary algebraic notions which are to replace the analytic notion of multiplier.

Step 2. Clarification of the structure of the endomorphism ring as far as necessary. In particular proving that this ring is $\mathbb{Z}$ or imaginary quadratic. In the latter case it is an order in an imaginary quadratic number field or, in some special cases, an order in a quaternion skew field.

Step 3. Conclusion: In the case of finite base field, definition of the Frobenius endomorphism π and using it to complete the proof of the RHp.

In the first two steps Hasse assumes that the base field K is algebraically closed. In the third step only he considers function fields with a finite base field; the results of the first two steps will then be applied to the base field extension with the algebraic closure of the base field. Note that the genus of a function field does not change under base field extensions—provided the new base field is separable over the original base field. (See Remark 4.4 on page 46.)

Let us begin with Step 1.

7.4.1 Meromorphisms and the Jacobian

Let $F|K$ be an elliptic function field with algebraically closed base field K of arbitrary characteristic. The task is to give an algebraic definition of the notion of *endomorphism ring* of the Jacobian of $F|K$. In the classical case, as I have mentioned in Sect. 7.3, such endomorphism was defined by means of a "multiplier", i.e., a complex number μ which, when multiplied with the lattice $\mathfrak{w}$ of periods, maps $\mathfrak{w}$ into itself: $\mu\mathfrak{w} \subset \mathfrak{w}$. This defines an endomorphism of the additive factor group $\mathbb{C}/\mathfrak{w}$, the latter being bijectively related to the points of the elliptic curve via the Weierstraß $\wp$-function (see (7.9), page 86). But in the algebraic environment there is no period lattice and hence no multiplier. One has to look for an algebraic analogue of the group $\mathbb{C}/\mathfrak{w}$ which is to carry the endomorphisms.

Hasse found it in the divisor class group $\mathcal{C}_0 = \mathcal{C}_0(F|K)$ of degree 0, i.e., the group of divisors of degree 0 of $F|K$ modulo principal divisors. The group operation in $\mathcal{C}_0$ is written here as multiplication whereas in $\mathbb{C}/\mathfrak{w}$ it was written as addition, but this is of no importance. There is a bijection of $\mathcal{C}_0$ with the set of primes P of $F|K$ as follows:

Fix a prime of $F|K$ and call it P_∞. Then, for every $C \in \mathcal{C}_0$ the divisor class $C \cdot P_\infty$ is of degree 1 and therefore, by the Riemann-Roch Theorem it contains exactly one integer divisor. This is necessarily a prime P. (Recall that here $F|K$ is elliptic, i.e., the genus is $g = 1$.) The map

$$C \mapsto P$$

is a bijection from the divisor classes $C \in \mathcal{C}_0$ to the primes P of $F|K$. It can be used to impose a group operation on the set of primes, written as addition, such that

$$\frac{P_1 + P_2}{P_\infty} \sim \frac{P_1}{P_\infty} \cdot \frac{P_2}{P_\infty} \tag{7.10}$$

(Recall that I use $\sim$ as the sign for divisor equivalence modulo principal divisors, see page 10). The chosen prime P_∞ acts as the zero element of that group.

The addition (7.10) can be explicitly described by means of coordinates, according to the well known addition formulas known from classical algebraic geometry. In the classical case the curve is usually given in Weierstraß normal form, and then the zero element of this addition is usually represented by the point at infinity (see page 86). This explains my notation P_∞ also in the abstract case. Hasse clearly preferred the abstract intrinsic definition (7.10) which does not refer to the coordinates of the respective points. In the introduction of his first paper for the elliptic case he wrote:

> *"In view of the problem to generalize the whole theory for arbitrary genus g I deliberately avoid to use special explicit formulas or knowledge about elliptic fields – even if the proofs may appear somewhat abstract for those who are only interested in the elliptic case. In particular I never need to use the explicit formulas for the Addition Theorem. Instead it is sufficient to look at the very roots of these formulas, namely the multiplication of divisor classes. The whole theory has now obtained a purely structural character."*

This did allow him to treat simultaneously the cases of all characteristics, including characteristic 2. In his eyes this was more adequate to the situation.

In geometrical language, what Hasse did was in fact the construction of the *Jacobian* $J = J(F|K)$. Since the genus is 1 the function field of the Jacobian of $F|K$ is isomorphic to the original field $F|K$. The K-rational points of J can be identified with the primes P of degree 1 of $F|K$, provided one of them is chosen to be the zero element. The group operation in J is written as addition, and J is essentially the same as the multiplicatively written group operation of C_0 in view of (7.10).

All this and also the following constructions were standard in classical algebraic geometry but in Hasse's time it was unknown whether and how they could be transferred to characteristic p. The work of Hasse in the 1930s on the RHp implied to a large extent the transfer of the classical algebraic geometry of curves from characteristic 0 to characteristic p.

But Hasse did not think in geometric terms. His mathematical background was algebraic number theory which he was familiar with (and to which he had given important outstanding contributions). His objects which he had worked with were number fields, rings, ideals and valuations. In particular he had successfully used p-adic number fields. He was convinced that these ideas are also helpful for the RHp—which after all turned out to be the case as we shall see. As far as I know he never anticipated to deal with problems of algebraic geometry as such. He did not foresee the merging of number theory and algebraic geometry to what today is called "arithmetic algebraic geometry". This was done by later generations of mathematicians.

After having defined the "Jacobian", Hasse's next task was to give an algebraic definition of "endomorphism" of the Jacobian in the framework of function fields.

To this end Hasse created the notion of "meromorphism" of an elliptic function field, defined as an isomorphism μ of $F|K$ into itself. In this connection he wrote μ as a right operator, thus for $x \in F$ its image is denoted by $x\mu$ and similarly for divisors, divisor classes etc. The image of F is $F\mu$. The degree $[F : F\mu]$ is finite. Consider the map for divisors A of $F|K$:

$$A \mapsto N_{F|F\mu}(A) \cdot \mu^{-1} \tag{7.11}$$

where $N_{F|F\mu}$ denotes the norm operator from F to its subfield $F\mu$.

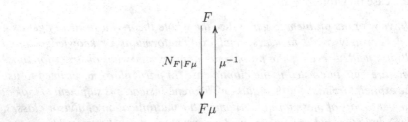

This defines an endomorphism of the divisor group of $F|K$, which respects principal divisors and divisor degrees. In particular we get an endomorphism of the divisor class group C_0 of degree 0.

Hasse denotes the map (7.11) also by μ but this time as operator from the left. Thus μA denotes the image of A under the map (7.11). This is a divisor of $F|K$ whereas $A\mu$ is a divisor of $F\mu|K$. This somewhat subtle notation, Hasse says, had been proposed by Witt. At that time Witt was in Göttingen and belonged to the small group of Hasse's seminar; he had already a reputation for his elegant style and notations. In Witt's notation, if P is a prime of $F|K$ then from (7.11) we have for $x \in F$

$$x\mu \cdot P = x \cdot \mu P \tag{7.12}$$

which says that the element $x\mu \in F\mu \subset F$ is mapped modulo P onto the same element as $x \in F$ modulo μP. This associativity allows to write $x\mu P$ without dot.

A meromorphism is said to be "normalized" with respect to P_∞ if $\mu P_\infty = P_\infty$. If this is the case then a glance at (7.10) convinces us that

$$\mu(P_1 + P_2) = \mu P_1 + \mu P_2. \tag{7.13}$$

In other words: If the meromorphism μ is normalized then it induces an endomorphism of the Jacobian J as an additive group. Every meromorphism can be normalized by multiplication with a suitable automorphism of $F|K$ which leaves C_0 elementwise fixed.

These "suitable automorphisms" are the translation automorphisms of $F|K$. The group of translation automorphisms is isomorphic to the divisor class group C_0 of degree 0. The translation τ_C belonging to the class $C \in C_0$ has the property that

$P\tau_C \sim PC$ for every prime P of J. If we write $C \sim A/P_\infty$ then $P\tau_C = P + A$ for all P. Nowadays the existence of those translation automorphisms is more or less evident since, as said above, $F|K$ can be viewed as the function field of the Jacobian which is an abelian variety. However, at that time the notion of abelian variety had not yet been established in characteristic p. Hence Hasse had to construct those automorphisms explicitly. In a letter dated 21 February 1933 his friend Davenport asked:

"Are you going to get new automorphisms or birational transformations from your method?"

And Hasse had to explain him that the translation automorphisms can be constructed as products of two reflections.

Hasse said that his idea of using the notion of meromorphism to define endomorphisms of the Jacobian came from the analogy to the classical case. (See page 86.) There, every multiplier μ defines a meromorphism of the function field $F = K(\wp(z), \wp'(z))$ by

$$\left(\wp(z), \wp'(z)\right) \;\mapsto\; \left(\wp(\mu z), \wp'(\mu z)\right). \tag{7.14}$$

One has to use the fact that every analytic function with period $\mathfrak{w}$ is a rational function of $\wp(z), \wp'(z)$. But in the algebraic case there do not exist multipliers and $\wp$-function to construct meromorphisms. Hence Hasse had to build his theory directly on the abstract algebraic notion of meromorphism as defined above.

Today the terminology of "meromorphism" is not in use any more. Instead, one says "isogeny" of an elliptic curve to itself. Nevertheless I shall use here Hasse's old terminology of "meromorphism". This terminology was new and unusual at that time, thus it will provide us a glimpse into the feeling of the times of the 1930s, namely that this is the beginning of some new development.

REMARK Looking at (7.14) we can understand how Hasse and Witt arrived at that "subtle" notation, as we have called it above. (See page 94.) If I denote the function $z \mapsto \wp(z)$ just by $\wp$ then the function $z \mapsto \wp(\mu z)$ should be written as $\wp\mu$, and similar for $\wp'$ and for all $f \in F = K(\wp, \wp')$. Thus in the classical case, the multiplicator $\mu : \mathfrak{w} \to \mathfrak{w}$ defines a meromorphism $f \mapsto f\mu$ of the function field, written as right operator. This led Hasse in the abstract case also to write the meromorphism $\mu : F \to F$ as right operator. And then the map $\mu : P \mapsto \mu P$ defined by the diagram on page 94 corresponds in the classical case to the multiplication $z \mapsto \mu z$ and therefore also in the abstract case was written as left operator.

As explained above, every meromorphism μ of $F|K$ (in the abstract sense) induces an endomorphism of the group C_0 of divisor classes of degree 0; if it is normalized then it induces an endomorphism of the Jacobian J, if J is understood as the additive group of primes of $F|K$ given by (7.10).

But there is more to do: these endomorphisms should constitute a ring. Given two meromorphisms μ_1, μ_2 of $F|K$ with corresponding endomorphisms of J it has to be verified that the endomorphisms $\mu_1 + \mu_2$ and $\mu_1\mu_2$ of J are also coming from meromorphisms of the function field.

This is evident for the product since the product of two meromorphisms is just the successive application of the two maps. For the sum it is somewhat more involved, although Hasse's way to deal with this problem looks quite natural and easy for us.

7.4.2 The Double Field

Geometrically speaking, if $F = K(\Gamma)$ is regarded as the function field of a curve Γ over K then Hasse considers the function field of Γ over a field E isomorphic to F, i.e., the field $\mathcal{F} = E(\Gamma)$. This is a function field of two variables over K, regarded as a tower of two function fields of one variable. Hasse calls $\mathcal{F}$ the "double field" (*Doppelkörper*) belonging to F. It can be realized as the independent compositum of $F|K$ with $E|K$:

$$\mathcal{F} := F \cdot E = \mathrm{Quot}(F \otimes_K E)$$

where $\mathrm{Quot}(\dots)$ denotes the field of quotients.

Geometrically, $\mathcal{F}$ can be viewed as the function field over K of the surface $\Gamma \times \Gamma$, the direct product of Γ with itself, where the graphs of endomorphisms are located. But Hasse preferred to work with function fields of transcendence degree 1; so he introduced his "double field" $\mathcal{F}$ and considered it as a base field extension of $F|K$ with E as the new base field. See the diagram below.

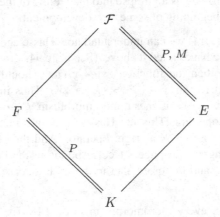

Every prime P of $F|K$ extends uniquely to a prime of the base field extension $\mathcal{F}|E$, and that extension will likewise be denoted by P. But there are more primes of $\mathcal{F}|E$, namely those for which the valuation is trivial on F. The residue map of such a prime induces an isomorphism of F into the residue field. These primes M are called "transcendental" because the image of F under the residue map is transcendental over K. The other primes P as above are then called "algebraic".

Let M be a transcendental prime and $\mathcal{F}M$ its residue field. Let $\mu : F \to \mathcal{F}M$ denote the isomorphism of F induced by the residue map modulo M. We have

$$\mathcal{F}M = F\mu \cdot E . \tag{7.15}$$

(The residue map of $\mathcal{F}|E$ modulo M is usually normalized such that the base field E is kept elementwise fixed.) This is the field compositum of the two fields $F\mu$ and E which are both transcendental over K. This field compositum (7.15) is not independent over K; its degree over each component $F\mu$ and E is finite. We have $[\mathcal{F}M : E] = \deg M$. In particular, if $\deg M = 1$ then the map $\mu : F \to E$, followed by a fixed K-isomorphism $E \xrightarrow{\approx} F$ is a meromorphism of $F|K$.

In this way Hasse interprets the meromorphisms of F as the transcendental primes of degree 1 of $\mathcal{F}|E$.

The study of the divisor class group of $\mathcal{F}|E$ then leads to the verification that, indeed, the endomorphisms of J as defined above form an additive group. In fact, if two meromorphisms μ_1, μ_2 are represented by the transcendental primes M_1, M_2 then, Hasse shows, their sum $\mu_1 + \mu_2$ is represented by the prime $M_1 + M_2$ given by

$$\frac{M_1 + M_2}{P_\infty} \sim \frac{M_1}{P_\infty} \cdot \frac{M_2}{P_\infty} \tag{7.16}$$

which is the same relation as in (7.10) but now to be read in the elliptic function field $\mathcal{F}|E$. (Well, if $\mu_1 + \mu_2 = 0$ then this is not represented by a transcendental prime but by the algebraic prime P_∞. In some sense the algebraic primes can be regarded as improper meromorphisms. In the following I will not always mention this exceptional case explicitly.)

Let us denote the ensuing ring by $\mathsf{M} = \mathsf{M}(F|K)$, similar as we did in the classical case. Hasse calls it the "ring of multipliers" of the elliptic function field $F|K$, in analogy to the terminology used in the classical theory of complex multiplication. But here let us use "ring of endomorphisms" of the Jacobian $J(F|K)$.

7.4.3 Norm Addition Formula

Hasse's abstract construction of the endomorphism ring M is neither difficult nor quite surprising to us, although it may have looked somewhat unfamiliar to his contemporaries. But the construction does not prove anything yet, it is only the first step, providing us with the algebraic framework in which the proof of the RHp can be executed.

Now I will turn to Step 2. (See page 91.)

Every meromorphism $\mu \in \mathsf{M}$ has a degree which Hasse called "norm":

$$\mathcal{N}(\mu) := [F : F\mu].$$

This name will be justified later only, for M will turn out to be (isomorphic to) a subring of an imaginary quadratic number field or skew field, in such a way that $\mathcal{N}(\mu)$ equals the norm of the algebraic number μ in this field. In view of this I have denoted Hasse's norm by the calligraphic sign $\mathcal{N}(\mu)$ as I had done above in the classic case. I put $\mathcal{N}(0) = 0$ and note that

$$\mathcal{N}(\mu\nu) = \mathcal{N}(\mu)\mathcal{N}(\nu) \tag{7.17}$$

since the product $\mu\nu$ is just the successive application of the meromorphisms μ and ν. Hasse's key result is the important

Norm Addition Theorem:

$$\mathcal{N}(\mu + \nu) + \mathcal{N}(\mu - \nu) = 2\mathcal{N}(\mu) + 2\mathcal{N}(\nu) \qquad \text{for } \mu, \nu \in \mathsf{M}. \tag{7.18}$$

Looking at this formula we get the impression that $\mathcal{N}(\mu)$ is a quadratic form on M. This can indeed be verified, the corresponding bilinear form being given as

$$(\mu, \nu) := \frac{1}{2}\Big(\mathcal{N}(\mu + \nu) - \mathcal{N}(\mu) - \mathcal{N}(\nu)\Big). \tag{7.19}$$

The bilinearity of this symbol is a consequence of the norm addition formula (7.18). From the product formula (7.17) we see that M has no zero divisors. Per induction one verifies from the norm addition formula that $\mathcal{N}(n) = n^2$ for $n \in \mathbb{Z}$. Hence M is of characteristic 0. For each $\mu \in \mathsf{M}$ consider the quadratic polynomial over $\mathbb{Z}$:

$$f(X) = X^2 - 2(\mu, 1)X + \mathcal{N}(\mu).$$

Using repeatedly the above mentioned properties of $\mathcal{N}(\ldots)$ one computes $\mathcal{N}(f(\mu)) = 0$, hence $f(\mu) = 0$, showing that either $\mu \in \mathbb{Z}$ or μ is a quadratic integer. Since $\mathcal{N}(\mu) > 0$ it follows that $\mathbb{Z}[\mu]$ is *imaginary* quadratic.

Hence:

Lemma *Let M be any ring with unit element 1, and assume that to every $\mu \in \mathsf{M}$ there is assigned a natural number $\mathcal{N}(\mu)$ such that (7.17) and (7.18) hold. Then either $\mathsf{M} = \mathbb{Z}$ and $\mathcal{N}(n) = n^2$ for every $n \in \mathbb{Z}$, or M is an order in an imaginary quadratic number field or quaternion algebra and $\mathcal{N}(\mu)$ is the norm of μ from this field. (In case of a quaternion algebra one has to take the reduced norm.)*

We see that, using this elementary lemma, the norm addition formula (7.18) leads quite formally to the imaginary quadratic structure of M and hence to the RHp for elliptic function fields (for this see the following Sect. 7.4.4).

Hasse talked about this lemma in his Göttingen seminar 1935. As he reports in his paper, some participants of the seminar provided a number of simplifications of his proof of the lemma. The first was Teichmüller. More was given by Hasse's student Behrbohm whose proof was published in Crelle's Journal 1935 [Beh35].[2] Witt refereed Behrbohm's article in the *Zentralblatt* and on that occasion provided further simplifications. Thus the final form of the proof of this lemma was the result of a teamwork of Hasse's seminar group in Göttingen 1935.

But how did Hasse get the idea for the norm addition formula, and how did he find its proof? We are in a good position to be able to follow this development by reading the letters which were frequently exchanged between Hasse and Davenport in those years. On 29 January 1934, the day before Hasse's departure for Hamburg to deliver his lecture series on the invitation from Artin, he had written to Davenport:

> *"I am very troubled at present because I found a gap in my proofs about the elliptic case while drawing up my lectures for Hamburg. The whole thing seems too sensible for being wrong. But it may be that the proof of the actual result lies a bit deeper than my argument went so far. The possibility I have to exclude is that the operation π is transcendental. I can prove that if it is algebraical it must be imaginary quadratic, because a unit operation cannot exist."*

(Hasse means that a unit operation of infinite order does not exist.)

Thus even at this stage, when Hasse was going to present in Hamburg a sketch of his new proof, he had not yet fully cleared up the structure of the meromorphism ring. He did not yet know the Norm Addition Theorem, he even did not know whether the meromorphism ring may contain transcendental elements. Even the Frobenius meromorphism π was not yet established as being algebraic over $\mathbb{Z}$. After all, π is the most important object in the RHp project (see the following Sect. 7.4.4). Fortunately Hasse was able to overcome this flaw, at least in the case of π, as he informed his friend shortly after his return to Marburg. On 12 February 1934 he wrote:

> *"Hamburg was a full success from every point of view. I was able to fill up the gap in my proof shortly after I wrote you how depressed I was. The new proof is, as Artin meant, even more adequate than the old would have been, were it consistent ..."*

---

[2]Hermann Behrbohm is mentioned in the list of Hasse's doctoral students, a list that was written down by Hasse himself. But finally Behrbohm switched to aircraft industry and did his Ph.D. in applied mathematics 1944. Behrbohm had only two papers in number theory: Besides of his above mentioned publication in the *Göttinger Nachrichten* there is one about the Euclidean algorithm in quadratic fields [BR36], jointly with the Hungarian mathematician L. Redei who in 1934/1935 was visiting Hasse in Göttingen as a Humboldt fellow.

Hasse continued his letter with the explanation of his new arguments which showed that π is algebraic, i.e., as an element in the endomorphism ring it satisfies an algebraic equation with integer coefficients. From this he could deduce that if $\pi \notin \mathbb{Z}$ then, using Dirichlet's Unit Theorem, the ring $\mathbb{Z}[\pi]$ is imaginary quadratic since it contains finitely many units only. For, if a normalized meromorphism induces a unit of M then it is necessarily an automorphism of the function field $F|K$, and normalization means that it leaves P_∞ fixed. It is well known that there are only finitely many such automorphisms of an elliptic function field. According to Dirichlet's Unit Theorem this can happen only if $\mathbb{Z}[\pi]$ is an order in an imaginary quadratic field (or $\pi \in \mathbb{Z}$).

But in the course of time Hasse abandoned this argument which is based on the Dirichlet Unit Theorem, in favour of the above Norm Addition Theorem.

The idea for the Norm Addition Theorem came about through Davenport. During the years 1934–1935 many of their letters which they exchanged contained remarks on the structure of elliptic function fields in characteristic p. Hasse often informed Davenport about his progress, and Davenport gave his comments and sometimes also his own proofs. Davenport's proofs were of quite different style as Hasse's since Davenport preferred explicit computations with coordinates whereas Hasse tried to use abstract arguments if possible. Davenport could never quite agree with Hasse's *dictum* that in mathematics not only the result is important but also its proof, and the way how that proof is embedded into a more comprehensive theory.

During August and September of 1935 there were no letters between Hasse and Davenport. For in that time Davenport stayed in Göttingen with the Hasses. (By the way, in spring time that year Hasse had been in England with Davenport.) Back in England, Davenport wrote to Hasse and sent his "heartiest thanks for the splendid time in Göttingen". This letter, dated 4 October 1935, started an exchange of particular high frequency: 20 letters until the end of the year and many more thereafter. The dominant theme was the proof of the RHp in the elliptic case.

Davenport informed Hasse that he had obtained the following formula for Hasse's norm:

$$\mathcal{N}(\mu + \nu) \le 2\mathcal{N}(\mu) + 2\mathcal{N}(\nu). \qquad (7.20)$$

He had obtained it in the course of proving the commutativity and algebraicity of the whole meromorphism ring M.(As to commutativity, Davenport's result turned out not to be correct and hence he had to modify it. Somewhat later Hasse discovered the "supersingular" fields, which are elliptic function fields of characteristic p with non-commutative ring of endomorphisms.)

In reply Hasse wrote on 21 November 1935 that this is a very remarkable progress. He had tried to obtain Davenport's inequality in his own way and on the way found out that it can be improved to an exact equality, viz. the norm addition formula. This, Hasse wrote, changed completely his approach since it immediately provided the positive definite quadratic form. When Hasse informed Davenport about it, the latter seems to have hesitated to submit his own proof for publication, with the argument that it had already been improved by Hasse. But Hasse protested; he wrote on 27 November 1935:

> *"My dear Harold, of course you must publish your proof! I have mentioned that you first had the idea of considering $\mathcal{N}(\mu)$ as a sort of absolute value and proved the algebraicity and commutativity of normalized meromorphisms on this basis, in both my preliminary paper for the Göttinger Nachrichten and my detailed account for Crelle's Journal..."*

Upon this Davenport decided to publish his paper, not in Crelle's Journal which Hasse had offered him but in the proceedings of the Cambridge Philosophical Society [Dav36]. He had motivated this choice already in an earlier letter dated 14 November 1935 as follows:

> *"As regards my proof of commutativity + algebraicity, I must say that I should like to publish it in England ... assuming it is O.K. Firstly, because it is definitely in my interest at present to publish as many moderately good papers in England as possible; secondly because I should not like to see it buried (if you will forgive the word) in a paper whose main emphasis will not be on new results but on new exposition. You know my prejudice that a new exposition is o(a new result)."*

(The small o refers to Landau's notation.)

This sounds quite harsh but it reflects the disagreement between Hasse and Davenport concerning basic views on the presentation of mathematics. Many years later, when Davenport was already established as one of the great masters of Number Theory, he used to say that although he had learned a great deal from Hasse, he had learned not nearly as much as he could have done if he had been "less pig-headed". This is reported by Halberstam in his comments to Davenport's Collected Works [Dav77].

Hasse himself said that he had got the idea of the Norm Addition Theorem and its proof from the classic theory of complex multiplication. For, there one knows how to find the poles and zeros of $\wp(\mu z) - \wp(\nu z)$ where $\wp(z)$ is the Weierstrass $\wp$-function. In the algebraic situation one has to replace $\wp(z)$ by an element x in the function field with pole divisor P_∞^2 and then find the prime decomposition of the element $x\mu - x\nu$ in F. It turns out that

$$x\mu - x\nu \cong \frac{P_\infty(\mu + \nu) \cdot P_\infty(\mu - \nu)}{(P_\infty\mu)^2 \cdot (P_\infty\nu)^2}. \tag{7.21}$$

(I am using Witt's notation as explained on page 94. In particular, if we write for brevity $\varrho = \mu + \nu$ then $P_\infty\varrho$ is a prime in the field $F\varrho$, it is the image of the prime P_∞ under the meromorphism ϱ. Since $F\varrho \subset F$ the divisors of $F\varrho$ can be considered as divisors of F (conorm). Remember that the symbol "$\cong$" indicates the decomposition of the element $x\mu - x\nu$ into divisors of $F|K$.)

Since the principal divisor on the left hand side has degree 0, the Norm Addition Theorem follows by comparing the degrees of numerator and denominator on the right hand side.

Hasse's algebraic proof of (7.21), valid in every characteristic, is given in [Has36c]. It is straightforward but somewhat lengthy since a number of special cases had to be treated separately. (For instance, when $\mu + \nu = 0$ or $\mu - \nu = 0$.)

The following result about the norm $\mathcal{N}(\mu)$ is almost obvious:

The norm $\mathcal{N}(\mu)$ equals the order of the kernel of μ as an endomorphism of the Jacobian – provided the algebraic field extension $F | F\mu$ is separable.

For, remember that $\mathcal{N}(\mu) = [F : F\mu]$. If $\mu(P) = P_0$ then by definition $N_{F|F\mu}(P) = P_0\mu$, which is to say that the prime P of F lies over the prime $P_0\mu$ of $F\mu$. Since the base field K is supposed to be algebraically closed, there are precisely $[F : F\mu]$ such primes P of F. Note that the separable extension $F|F\mu$ is unramified since both fields have genus 1.

REMARK A little closer look into the situation shows that the Galois group of $F|F\mu$ consists of the translation automorphisms τ_P where P belongs to the kernel of μ. If $F|F\mu$ is not separable then the same arguments apply to the maximal separable intermediate extension. Hence the above statement remains true if $\mathcal{N}(\mu)$ is replaced by its separable part $\mathcal{N}_{\text{sep}}(\mu) := [F : F\mu]_{\text{sep}}$.

7.4.4 The Frobenius Operator

After developing all this machinery, Hasse finally comes to the point, i.e., to the proof of RHp for elliptic function fields.

The RHp is concerned with function fields over a finite base field. But in the foregoing sections it was assumed that the base field K is algebraically closed. Since I do not want to change notations at this point, let me denote by $F_q | K_q$ a function field with finite base field K_q of q elements. Let K denote the algebraic closure of K_q and $F = F_q K$ the corresponding base field extension of $F_q | K_q$. Note that the genus of $F_q | K_q$ is the same as the genus of $F | K$.

Every prime P of $F | K$ is of degree 1. It induces in F_q a prime with residue field $F_q P$. The degree of that prime is $[F_q P : K]$. The RHp to be proved is concerned with the primes of degree 1 of F_q, i.e., for which $F_q P = K$. We have denoted the number of these primes by N.

One of Hasse's basic discoveries was what today is called the *Frobenius operator* belonging to F_q. Actually, the word "Frobenius" in this context was introduced later only, perhaps by André Weil? Hasse used to say simply "meromorphism π". Its definition is as follows:

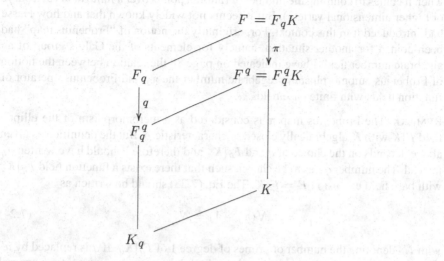

Since F_q and K are linearly disjoint over K_q there is one and only one meromorphism π of $F|K$ which in F_q induces the map $x \mapsto x^q$. (Note that raising to the power q leaves the elements of K_q fixed.) We have $F\pi = F_q^q K = F^q$ and hence

$$\mathcal{N}(\pi) = [F : F^q] = [F_q : F_q^q] = q .$$

Let $x \in F_q$. We compute, using Witt's notation (see (7.12), page 94):

$$x\pi P = x^q P = (xP)^q \begin{cases} = xP \text{ if } xP \in K_q \\ \neq xP \text{ if } xP \notin K_q \end{cases} \tag{7.22}$$

(provided $xP \neq \infty$). We conclude that $\pi P = P$ if and only if $F_q P = K_q$, i.e., if the induced prime of P in F_q is of degree 1. Thus, the number N of primes of degree 1 of F_q equals the number of primes P of F which are fixed by π, i.e. which are in the kernel of $\pi - 1$. Hence, using what I have said on page 102 we conclude

$$\mathcal{N}(\pi - 1) = N \tag{7.23}$$

since $\pi - 1$ is separable.

Recall that $\pi \in M$ is contained in an imaginary quadratic number field due to the norm addition formula (see the lemma on page 98). This solves the problem (7.4) on page 84. In this way Hasse had completely "knocked the bottom" out of the RHp for elliptic fields, if I am allowed to use the words of Mordell in his letter of 9 March 1933.

Today the notion of Frobenius operator is well established in arithmetic geometry when it comes to counting the number of rational points over a finite field (of a curve or higher dimensional variety). But it seems not widely known that and how Hasse had introduced it in this context. For, originally the notion of "Frobenius map" had been coined for another situation, namely for elements of the Galois group of an algebraic number field. I have indicated on page 88 the relation between the notion of Frobenius automorphism in algebraic number theory and Frobenius operator of function fields with finite base fields.

REMARK The Frobenius map π is considered as a meromorphism of the elliptic field $F|K$ with K algebraically closed of characteristic p. But the definition as given above depends on the choice of q and $F_q|K_q$ and therefore I should have written π_q instead. The number q has to be chosen such that there exists a function field $F_q|K_q$ with base field extension $F = F_q K$. The Eq. (7.23) should be written as

$$\mathcal{N}(\pi_q - 1) = N_q \tag{7.24}$$

with N_q denoting the number of primes of degree 1 of $F_q|K_q$. If q is replaced by q^r and F_q by $F_{q^r} = F_q K_{q^r}$ then $\pi_{q^r} = \pi_q^r$ which is in accordance to Artin's general observation for arbitrary genus (see page 51).

The smallest q which can be chosen in this way is the number of elements of the field $\mathbb{F}_p(j)$ where $j = j_{F|K}$ denotes the absolute invariant of the elliptic field $F|K$. (For a discussion about invariants see Sect. 8.4.1.)

7.5 Some Comments

7.5.1 Rosati's Anti-automorphism

The concept and proof of Hasse's norm addition formula (7.18) was a formidable task, considering that previously nothing was known about the arithmetical structure of elliptic function fields in characteristic p. It constituted a fundamental step in Hasse's proof the RHp in the elliptic case.

However, as we observe often also in other situations of mathematics, in the course of later development the norm addition formula lost its importance and was reduced to a side remark, at least for the proof of RHp in the elliptic case. This is the consequence of the work of Deuring in his second paper on correspondences [Deu40]. Deuring had discovered the algebraic version of Rosati's anti-automorphism $\mu \mapsto \mu'$ of the ring M of endomorphisms. This is obtained by interchanging F and E in the double field, see Sect. 7.4.2. In the elliptic case this satisfies

$$\mu\mu' = \mu'\mu = \mathcal{N}(\mu).$$

Since $\mathcal{N}(\mu) > 0$ if $\mu \neq 0$, we see that $\mu \mapsto \mu\mu'$ is a positive definite quadratic form on M. In particular $\pi\pi' = q$ and $(\pi-1)(\pi-1)' = (\pi-1)(\pi'-1) = \mathcal{N}(\pi-1) = N$.

Actually, already in 1934 Hasse had a similar idea, just before his lecture in Hamburg in January. There is a letter of Hasse to Davenport dated 12 February 1934 where he explains in detail his ideas for his new proof. There he constructed a meromorphism π' with

$$\pi\pi' = q \quad \text{and} \quad \pi + \pi' = q + 1 - N.$$

This however worked only in characteristic $\neq 2, 3$ since he heavily relied on explicit and somewhat involved computations with Weierstrass normal form. Perhaps his wish to obtain a complete and more lucid proof of what he called "complex multiplication in characteristic p" induced him not to stop already in 1934 with the elliptic case. Surely he also wished to include characteristics 2 and 3.

Summary

Parallel to his joint work with Davenport which I have discussed in the foregoing chapter, Hasse investigated elliptic function fields $F|K$ with finite base fields K, with the aim of proving the RHp. He obtained his first proof in early March 1933. For an elliptic field (or, equivalently, an elliptic curve) the genus is 1 and hence F.K. Schmidt's polynomial $L^*(t)$ is quadratic. Artin's criterion requires that its discriminant is ≤ 0, i.e., its roots are imaginary quadratic. Hasse's first proof proceeds by lifting the given elliptic curve Γ over K to a suitable elliptic curve Γ^* defined over an algebraic number field and admitting complex multiplication. By carefully controlling the lifting process Hasse could achieve that the ring of multipliers of Γ^* (which is a subring of an imaginary quadratic field by the classical theory of complex multiplication) contains an element π, such that its quadratic equation equals F.K. Schmidt's polynomial of the original function field $F = K(\Gamma)$. The proof uses class field theory within the theory of complex multiplication, in particular Artin's reciprocity law.

This first proof of Hasse was never published but preliminary announcements appeared. From several documents of Hasse's legacy the proof can be recovered in all details. It is, however, not quite complete since for technical reasons Hasse had to exclude some cases which he could not cover in the first attempt. While writing up the manuscript for publication he tried to eliminate those restrictions and during this activity he found a simpler proof, valid quite generally, and working without detour over characteristic zero.

For the new proof Hasse had to algebraize the theory of complex multiplication, such as to become applicable to the case of finite base fields. He defines the "multipliers" of an elliptic function field by what today is called "isogeny" (but he uses the word "meromorphism") of the elliptic function field. These form naturally a ring without zero divisors and of characteristic 0. Hasse succeeds to find within this ring an element π, the so-called "Frobenius multiplier" which is a zero of F.K. Schmidt's quadratic polynomial. The crucial theorem is that this ring carries a positive definite quadratic form, showing that π is imaginary quadratic. In later years Deuring could simplify Hasse's proof by using the so-called Rosati anti-automorphism of the ring of multipliers, which he had defined algebraically for arbitrary function fields.

Chapter 8
More on Elliptic Fields

While Hasse worked on the RHp for elliptic fields he and his team discovered several properties of elliptic fields in characteristic p which were new and important although they were not absolutely necessary for the proof of RHp. In this chapter I am going to discuss some of them.

8.1 The Hasse Invariant A

I have already mentioned the Hasse invariant A of an elliptic function field $F|K$ at former occasions. (See pages 84, 87.) Now here is Hasse's definition:

Consider an elliptic function field $F|K$ of characteristic $p > 0$. It is assumed that the base field K is algebraically closed. Let ω denote the integer differential of $F|K$ ("differential of the first kind").

REMARK It just occurs to me that I have not yet discussed the notion of "differential" in a function field. Today this belongs to the standard prerequisites of the theory of function fields, and hence I may be allowed to use it freely here. But in Hasse's time the differentials were not yet firmly established as algebraic objects. Hence Hasse had to write another paper where he defined differentials algebraically and proved their relevant properties [Has34d]. In particular he defined the local expansions of a differential, the divisor of a differential (which belongs to the canonical class), the residue of a differential at a prime P, and he proved the Residue Theorem which says that the sum of these P-adic residues of a differential, for all P, vanishes (Cauchy's theorem). The integer differentials form a K-space of dimension g.

For an elliptic field $F|K$ we have $g = 1$ and hence the integer differential ω is unique up to a nonzero constant factor. ω has no zeros or poles. Choose a prime divisor P of $F|K$. It does not matter which one since all primes are conjugate under

© Springer Nature Switzerland AG 2018
P. Roquette, *The Riemann Hypothesis in Characteristic* p
in Historical Perspective, Lecture Notes in Mathematics 2222,
https://doi.org/10.1007/978-3-319-99067-5_8

the automorphism group of translations. For instance one can choose P as the prime P_∞ which acts as zero element of the Jacobian (see page 92). Consider the P-adic development

$$\omega = (c_0 + c_1 t + c_2 t^2 + \cdots) \, dt \qquad (8.1)$$

with respect to a prime element t for P. The coefficients c_i are in the base field K, and $c_0 \neq 0$ since ω has no zero. If the base field would be of characteristic 0 one could integrate

$$\int \omega = c_0 t + \frac{c_1}{2} t^2 + \frac{c_2}{3} t^3 + \cdots =: u$$

so that $\omega = du$ becomes an exact differential (locally at P). But if the base field K is of characteristic $p > 0$ this integration process is not possible because of the appearance of p in the denominators of coefficients. Hasse uses a certain approximation. He takes the following prime element u for P:

$$u := c_0 t + \frac{c_1}{2} t^2 + \frac{c_2}{3} t^3 + \cdots + \frac{c_{p-2}}{p-1} t^{p-1}$$

i.e., he integrates as far as it is possible in characteristic p. Then

$$\omega \equiv du \qquad\qquad \mod P^{p-1} \qquad\qquad (8.2)$$

$$\omega \equiv (1 + A u^{p-1}) \, du \mod P^p \qquad\qquad \text{with } A = \frac{c_{p-1}}{c_0^p} \in K . \qquad (8.3)$$

These congruences between differentials are to be understood locally, as differentials in the P-adic completion of F. If u' is another prime element for P such that $u' \equiv u \mod P^p$ then one obtains the same constant A. If ω is replaced by $c\,\omega$ with $0 \neq c \in K$ then A is replaced by $c^{-(p-1)} A$. Thus A is not really an invariant of the elliptic field but it is "*homogeneous of dimension* $-(p-1)$" as Hasse says. Since here K is assumed to be algebraically closed we see that A can be normalized such that $A = 1$ or $A = 0$.

After dividing by u^p it is seen from (8.3) that A appears as the coefficient of u^{-1} in the expansion of the differential $\frac{\omega}{u^p}$. Thus A is its residue:

$$A = \mathrm{res}_P \left(\frac{\omega}{u^p} \right) . \qquad\qquad (8.4)$$

In terms of an arbitrary prime element t at P this can be written as

$$A = \frac{1}{c_0^p} \, \mathrm{res}_P \left(\frac{\omega}{t^p} \right) \qquad\qquad (8.5)$$

This is the original definition of A by Hasse in [Has34b]. I have found it already in one of his letters to Davenport which is dated 5 November 1933. It turned out to be important since it governs the structure of the endomorphism ring of the Jacobian in characteristic p. It has no analogue in characteristic 0. In my opinion its discovery was a historic achievement.

But where and how did Hasse discover this invariant?

8.2 Unramified Cyclic Extensions of Degree p

In his paper [Has34b] Hasse studied unramified cyclic extensions of degree p of an elliptic function field. On this occasion he found:

Theorem A *Let $F|K$ be an elliptic function field of characteristic p with algebraically closed base field. There exists an unramified cyclic extension $F'|F$ of degree p if and only if Hasse's invariant $A \neq 0$. If this is the case then $F'|F$ is unique.*

The proof is simple and straightforward. Nevertheless I believe it is worthwhile to have a look at it.

The starting point of the proof is the observation that a cyclic extension $F'|F$ of degree p is generated by an Artin-Schreier equation

$$F' = F(y), \qquad y^p - y = z \in F, \qquad (8.6)$$

as shown by Artin and Schreier [AS27a]. The element z in (8.6) is uniquely determined up to a transformation

$$z \to z + (s^p - s) \qquad \text{with } s \in F. \qquad (8.7)$$

The second observation is that a prime P of F is unramified in F' if and only if s in (8.7) can be chosen such that $v_P(z + s^p - s) \geq 0$. If P is a pole of z then this implies that $v_P(z) \equiv 0 \bmod p$. This was one result in Hasse's earlier paper [Has34e] on the arithmetic behavior of Artin-Schreier extensions. I have already mentioned that paper in Sect. 6.3.3 on page 74.

Let $P_1, \ldots P_n$ denote the poles of z. If $F'|F$ is unramified then for each P_i there exists $s_i \in F$ with $v_{P_i}(z + s_i^P - s_i) \geq 0$. By approximating the s_i simultaneously we find $s \in F$ such that $v_{P_i}(z + s^p - s) \geq 0$ for all $i = 1, \ldots, n$. We may also apply the Strong Approximation Theorem ("Chinese Remainder Theorem") to the Dedekind ring which is the intersection of all the valuation rings $\mathcal{O}_{P'}$ for $P' \neq P$, and find s such that $v_{P'}(z + s^p - s) \geq 0$ for all P' different from one fixed prime P chosen in advance. In other words: After suitable choice we may assume that z in (8.6) has P as its only pole.

Let $v_P(z) = -r$ so that $z \in \mathcal{L}(P^r)$. Again, P is unramified and hence $r \equiv 0 \bmod p$. If $r > p$ consider $r/p > 1$. By the Riemann-Roch Theorem $\mathcal{L}(P^{r/p})$ is of

K-dimension r/p and there exists $s \in \mathcal{L}(P^{r/p})$ with the precise denominator $P^{r/p}$. If s is suitably normalized by a factor from K then we have that $z + s^p - s$ has pole order $< r$. Replacing z by $z + s^p - s$ and repeating this process we may finally assume that $r = p$, i.e., $v_P(z) = -p$. Conclusion:

Every unramified cyclic extension $F'|F$ of degree p is generated by an Artin-Schreier equation (8.6) where z has only one pole P which may be fixed in advance, and $v_P(z) = -p$. Moreover, there exists $s \in F$ such that

$$v_P(z + s^p - s) \geq 0. \tag{8.8}$$

This implies $v_P(s) = -1$, hence $s = \frac{c}{u} + \cdots$ where $0 \neq c \in K$ and u is the P-adic prime element appearing in the definition of A above. We now have

$$z + \frac{c^p}{u^p} - \frac{c}{u} = a \qquad \text{with} \quad v_P(a) \geq 0.$$

Multiplying with the integer differential ω and taking residues gives:

$$\mathrm{res}_P(\omega z) + c^p \, \mathrm{res}_P\left(\frac{\omega}{u^p}\right) - c \, \mathrm{res}_P\left(\frac{\omega}{u}\right) = \mathrm{res}_P(\omega a).$$

Here,

$$\mathrm{res}_P(\omega z) = 0 \ \text{ by the Residue Theorem for differentials,}$$

$$\mathrm{res}_P\left(\frac{\omega}{u^p}\right) = A \ \text{ by definition of } A, \text{ see (8.4),}$$

$$\mathrm{res}_P\left(\frac{\omega}{u}\right) = 1 \ \text{ by definition of residue, see (8.3),}$$

$$\mathrm{res}_P(\omega a) = 0 \ \text{ since } v_P(\omega a) \geq 0.$$

It follows $c^p A - c = 0$, hence $A = c^{-(p-1)} \neq 0$.

Conversely, assume $A \neq 0$. Fix a prime P of $F|K$. Consider the module $\mathcal{L}(P^p)$ in the sense of the Riemann-Roch Theorem. Since $F|K$ is elliptic, the dimension of $\mathcal{L}(P^p)$ is p. In fact, $\dim \mathcal{L}(P^i) = i$ for $i \geq 2$. Choosing the prime element u as above, every element $z \in \mathcal{L}(P^p)$ admits an expansion of the form

$$z \equiv \frac{c_p}{u^p} + \frac{c_{p-1}}{u^{p-1}} + \cdots + \frac{c_2}{u^2} + \frac{c_1}{u} \mod \mathcal{O}_P$$

with coefficients $c_i \in K$. ($\mathcal{O}_P$ denotes the valuation ring of P.) If $v_P(z) = -i$ then $c_i \neq 0$ while $c_j = 0$ for $j > i$. Disregarding the last term $\frac{c_1}{u}$ for a moment, that expansion defines a surjection of K-modules

$$\mathcal{L}(P^p) \longrightarrow \sum_{0 \leq i \leq p-2} K \frac{1}{u^{p-i}}$$

with kernel K. Hence there exists $z \in \mathcal{L}(P^p)$, uniquely determined modulo K, which is the preimage of $\frac{1}{u^p}$. That is, z has the expansion

$$z \equiv \frac{1}{u^p} + \frac{c_1}{u} \bmod \mathcal{O}_P \qquad (8.9)$$

with $c_1 \in K$. Multiplying by the differential ω and taking residues as above yields

$$0 = A + c_1. \qquad (8.10)$$

This time we know that $A \neq 0$, hence $c_1 \neq 0$. Choose $b \in K$ such that $b^{p-1} = -c_1^{-1}$. Then (8.9) can be written in the form

$$b^p z \equiv -s^p + s \bmod \mathcal{O}_P \qquad \text{with } s = -\frac{b}{u}.$$

Replacing z by $b^p z$ we see that (8.8) holds.

The uniqueness of the unramified cyclic extension $F'|F$ is seen by inspection of the above proof which yields the element z unique modulo K.

8.2.1 The Hasse-Witt Matrix

Although this Chap. 8 is devoted predominantly to elliptic function fields let me briefly report here about the generalization of Theorem A to function fields of higher genus. This is contained in the joint paper of Hasse and Witt [HW36].

First I have to give the definition of A of a function field $F|K$ of genus $g > 0$. This time A will be a $g \times g$ matrix with coefficients in K. For simplicity the base field K is assumed to be algebraically closed. Let $\omega = (\omega_1, \ldots, \omega_g)$ be a basis of the integer differentials of $F|K$. Let $P_1, \ldots, P_g$ be different prime divisors such that their product $A = (P_1 P_2 \cdots P_g)$ has $\dim(A) = 1$. Such divisor is called "non-special". The existence of a non-special divisor can be proved as a consequence of the Riemann-Roch Theorem, provided the base field is infinite. (If not then there exists a finite base field extension which carries a non-special divisor.) Developing each ω_i with respect to a P_j-adic prime element t_j in the form:

$$\omega_i \equiv \sum_{0 \leq \nu \leq p-1} b_{i,j}^{(\nu)} t_j^{\nu} dt_j \bmod P_j^p,$$

one obtains $g \times g$ matrices $B_\nu = (b_{i,j}^{(\nu)})$ with coefficients in K.

Definition : $A = B_0^{-1} B_{p-1}.$

(Compare with (8.3) for $g = 1$). This matrix A is unique up to the substitution

$$A \longrightarrow SAS^{-p}$$

with a regular matrix S over K. In this connection S^p denotes the matrix whose entries are the p-th powers of the entries of S (and not the p-th matrix power).

A is called the "Hasse-Witt matrix" of the function field $F|K$. The main result of Hasse and Witt is

> **Theorem A for Arbitrary Genus:** *Let $F|K$ be a function field of characteristic p of genus $g > 0$ with algebraically closed base field, and A its Hasse-Witt $g \times g$-matrix. The number of independent cyclic unramified extensions of F equals the rank γ of the matrix $AA^p A^{p^2} \cdots A^{p^{g-1}}$.*

In particular it follows $\gamma \leq g$, and $\gamma = g$ if and only if $\det A \neq 0$.

The Hasse-Witt matrix A has been later investigated in more detail, e.g., by J. I. Manin [Man65].

8.3   Group Structure of the Jacobian

Hasse's main idea for his second proof of RHp was the algebraic definition and investigation of the endomorphism ring of the Jacobian of a function field $F|K$. In this connection it is understandable that one of the first questions which came up was about the group structure of the Jacobian group J itself—although in retrospective it turned out that this is not really necessary for the proof of the RHp.

Suppose for simplicity that the base field K is algebraically closed.

Algebraically the Jacobian J of a function field $F|K$ is defined as the group C_0 of divisor classes of degree 0.

In the framework of algebraic geometry J carries a unique structure of abelian variety whose dimension is g, the genus of the function field. But in the present context it will suffice to consider J as a group only.

Consider a function field $F|K$ of genus $g = 1$, i.e., $F|K$ is elliptic. The base field is assumed to be algebraically closed. Let J_{fin} be the group of the elements in the Jacobian which are of finite order. (If the base field K is the algebraic closure of a finite field then $J_{\text{fin}} = J$.) First note that J and hence J_{fin} is a *divisible* group. In fact, for every meromorphism $\mu : F \to F\mu \subset F$ the induced map $\mu : J \to J$ is surjective with finite kernel. (Compare the definition (7.11) on page 94.)

Let h_n denote the number of the elements in J_{fin} whose order divides n. In the classical case when $K = \mathbb{C}$ is the complex number field it is known that

$$h_n = n^2 \tag{8.11}$$

for all natural numbers n. The exponent 2 is due to the fact that the periods of integrals form a 2-dimensional lattice $\mathfrak{w}$ which in turn mirrors the homology of the corresponding Riemann surface of genus 1. The Jacobian group J is isomorphic to the factor group $\mathbb{C}/\mathfrak{w}$ which is isomorphic to $\mathbb{R}/\mathbb{Z} \times \mathbb{R}/\mathbb{Z}$ (see Sect. 7.3). This implies (8.11) for all n.

Does (8.11) also hold for elliptic function fields $F|K$ of characteristic $p > 0$?

Hasse had early investigated this question. On 6 October 1933 he wrote to Davenport:

"*I have proved that the automorphism group given by the Addition Theorem is of 'dimension 2'.*"

Here Hasse refers to the group of translation automorphisms of the elliptic field $F|K$. (See page 94.) By definition this group is isomorphic to the Jacobian J of the function field. Hasse puts 'dimension 2' in quotes in order to indicate something like (8.11), but not quite the same since the prime number p (the characteristic) behaves differently than the other primes. Hasse writes that

"*J_{fin} is isomorphic to the additive group of all pairs (r_1, r_2) of rational numbers with denominator prime to p, considered mod 1.*"

In other words: He claims that (8.11) holds for those n which are prime to the characteristic p, whereas $h_n = 1$ if n is a power of p. This statement however, written on 6 October 1933, turned out not to be correct with respect to the contribution of the characteristic p. Soon Hasse discovered that there do exist elliptic function fields $F|K$, the Jacobians of which admit non-trivial elements of p-power order. He explains this to Davenport in his letter of 5 November 1933. He gives to Davenport the definition of the invariant A in the way I have done it in Sect. 8.1. And Hasse remarks that for $q = 5$ and $F = K(x, y)$ given by the Weierstrass normal form

$$y^2 = 4x^3 - g_2 x - g_3 \qquad \text{with} \qquad \omega = \frac{\mathrm{d}x}{y} \qquad (8.12)$$

one has $A = g_2$. Hence both cases $A = 0$ and $A \neq 0$ are possible. In the published paper [Has36c] he states the correct general result:

$$h_n = \begin{cases} n^2 & \text{if } n \neq 0 \bmod p \\ n & \text{if } n = p^r \text{ and } A \neq 0 \\ 1 & \text{if } n = p^r \text{ and } A = 0 \end{cases} \qquad (8.13)$$

Thus again, the Hasse invariant A comes into the play, it governs the group structure of the p-part of the Jacobian.

The relations (8.13) are an immediate consequence of Hasse's norm addition formula (Sect. 7.4.3). For, let μ_n denote the meromorphism inducing in J the multiplication by n. The norm addition formula implies $[F : F\mu_n] = n^2$. If $n \not\equiv 0 \bmod p$ then $F|F\mu_n$ is separable and therefore $[F : F\mu_n]$ equals the order of the kernel of μ_n in J, i.e., $n^2 = h_n$ (see page 102).

If n is a power of p then h_n equals the separability degree of $F|F\mu_n$ (see page 102). It suffices to discuss the case $n = p$ since then (8.13) for $n = p^r$ follows by induction. (Recall that C_0 is divisible.) We have to use Hasse's Theorem A of the foregoing section. $F\mu_p$ is isomorphic to F and hence there exists at most one cyclic unramified subextension of $F|F\mu_p$ of degree p. This implies that $F|F\mu_p$ is inseparable. If the separability degree is p then there exists a cyclic unramified subextension of $F|F\mu_p$ and hence $A \neq 0$ by Theorem A. And conversely.

The statement (8.13) is to be found in [Has36c] which contains Hasse's final proof of the RHp. That paper is divided into three parts with the following title and subtitles:

On the theory of abstract elliptic function fields.

I. *The structure of the divisor classes of finite order.*
II. *Automorphisms and meromorphisms. The Norm Addition Theorem.*
III. *The structure of the ring of meromorphisms. The Riemann hypothesis.*

The statement (8.13) and a proof of it are contained in Part I. But Hasse's proof which I have presented above uses the Norm Addition Theorem, and that comes later in Hasse's trilogy, namely in Part III. So why did Hasse publish two different proofs, one in Part I and a second one in Part III ? The following words in the introduction to Part I provide an explanation. Hasse wrote there:

> *"At present I am not sure whether the proof in this first part, based mainly on computations, can perhaps be entirely omitted. But for the time being I cannot do without it in certain arguments in the next parts."*

This suggests that at the time when Hasse had finished Part I the other parts were not yet complete. In fact, the manuscript for Part I had been submitted to Crelle's Journal on 16 October 1935 whereas Hasse had discovered his norm addition formula later, namely on 20. November 1935 according to his letter to Davenport the following day. (See page 100.) Hasse wrote that he had found the norm addition formula *"to my own great surprise."*

It appears that Hasse quickly rewrote his Part III and based it on the Norm Addition Theorem but left Part I unchanged since it had been submitted already. The manuscript for Part III was submitted on 22 November 1935—2 days after Hasse had found the norm addition formula. This hurry may be explained in part by the fact that Hasse wanted his proof of the RHp to appear prior to the International Congress in Oslo which was scheduled in September 1936. For he had been invited to give an 1-h talk in Oslo about his results on the RHp.

Thus the proof of (8.13) given in Part I is to be considered as being superseded by application of the norm addition formula.

REMARK For function fields $F|K$ of arbitrary genus $g \geq 1$ the formula (8.13) reads:

$$h_n = \begin{cases} n^{2g} & \text{if } n \not\equiv 0 \bmod p \\ n^g & \text{if } n = p^r \text{ and the Hasse-Witt matrix } A \text{ is regular} \\ n^\gamma & \text{with } \gamma < g \text{ else.} \end{cases} \qquad (8.14)$$

Here $\gamma \leq g$ is determined by means of the Hasse-Witt matrix A as described in Sect. 8.2.1. The case $n = p^r$ in (8.14) follows (by class field theory) from the result of Hasse-Witt and was known at least since the appearance of their paper [HW36]. However the case $n \not\equiv 0 \bmod p$ seems not to have been known yet at the time. Of course it was well known in the classical case due to the analytic theory of integrals and their periods. But an algebraic proof which would apply for every characteristic was not yet available. Hasse suspected its validity and I am quite sure that he tried to find a way how to prove it since he had settled it for the case $g = 1$. Today we could lift the function field $F|K$ to characteristic 0 such that it becomes a good reduction in the sense of Deuring. Then we have to observe that the divisor classes of order $\not\equiv 0 \bmod p$ are mapped injectively under good reduction. But this argument had not been available at the time. Perhaps Hasse finally did not care too much about it since his experience in the elliptic case had shown that the RHp could be proved without knowing the group structure of the Jacobian. The essential feature is the structure of the *ring of endomorphisms of the Jacobian* which, as he had shown in the elliptic case, carries a positive definite quadratic form given by the "Norm Addition Theorem" (page 98). Now Hasse's investigations concentrated on constructing such a positive definite form also in the case of higher genus (see page 148).

8.3.1 Higher Derivations

I have said that Hasse's proof of (8.13) had been superseded by the Norm Addition Theorem. Nevertheless the *method* developed by Hasse for the proof in Part I is of historic interest. Namely, Hasse worked with so-called "differential determinants". Such determinants had been commonly used in the classical case in characteristic 0. (Hasse refers to the book by Hensel and Landsberg [HL02].) But they do not work in characteristic p. Hasse had to modify the theory such as to be applicable also in characteristic p. To this end he wrote a separate paper about higher derivations in characteristic p [Has36a]. The date of submission of that paper is given as 15. October 1935—1 day before he submitted Part I.

Let $F|K$ be a function field. An additive K-map $D : F \to F$ is called a derivation if

$$D(yz) = yD(z) + D(y)z. \qquad (8.15)$$

In characteristic 0 one iterates. Let D^i denote the i-th iteration of D. If $M \subset F$ is a K-module with basis $y_0, \ldots, y_{n-1}$ the differential determinant

$$\Delta(M) := \det \begin{vmatrix} D^0(y_0) & D^1(y_0) & \cdots & D^{n-1}(y_0) \\ D^0(y_1) & D^1(y_1) & \cdots & D^{n-1}(y_1) \\ \cdots & \cdots & \cdots & \cdots \\ D^0(y_{n-1}) & D^1(y_{n-1}) & \cdots & D^{n-1}(y_{n-1}) \end{vmatrix} \quad (8.16)$$

is sometimes called the "Wronski determinant" of M. It is independent of the choice of basis up to a constant factor $c \in K$.

But in characteristic $p > 0$ this is not much useful since the p-th iteration D^p annihilates every power y^k with $k \geq p$. Therefore Hasse introduced a modified notion of higher derivation. Hasse replaced the i-th iteration D^i by $\frac{1}{i!} D^i$. Changing notation and calling this again D^i the determinant (8.16) with the new D^i turns out to be useful also in characteristic $p > 0$. More precisely, a system D^i of K-linear maps $F \to F$ $(i = 0, 1, 2, \ldots)$ with $D^0(x) = x$ is called a "higher derivation" if

$$D^i(xy) = \sum_{0 \leq j \leq i} D^j(x) D^{i-j}(y) \quad (8.17)$$

$$D^i(x) = \binom{i}{j} D^j D^{i-j}(x) \quad \text{if } j \leq i. \quad (8.18)$$

(Nowadays this is called a "hyperderivation".) Formula (8.17) is the analogue for the new D^i of the well known "Leibniz rule" for the $\frac{1}{i!} D^i$ in characteristic 0. Formula (8.18) is just the analogue of the iteration rule for the old D^i. Of course, it is required in addition that the elements c in the base field K are "differential constants" in the sense that $D^i(c) = 0$ for all $i > 0$.

If $x \in F$ is a separating element then there exists one and only one higher derivation $D_x^i(y)$ of $F|K$ (in the new sense) such that

$$D_x^1(x) = 1 \quad \text{and} \quad D_x^i(x) = 0 \quad \text{for} \quad i > 1.$$

This is the analogue to $\frac{1}{i!} \frac{d^i y}{dx^i}$ in characteristic 0. If t is another separating element then one can express $D_t^i(y)$ by the $D_x^j(y)$ and the $D_t^j(x)$ for $(j \leq i)$. This is called the "chain rule". I do not write down the explicit formulas for the chain rule; they are obtained from the ordinary chain rule in characteristic 0 by replacing the operators $\frac{1}{i!} \frac{d^i}{dx^i}$ by the new D_x^i and similarly for D_t^j.

The essential fact which Hasse had observed is that with these differential determinants (8.16) in the new sense one can work quite in a similar way as with the Wronski determinants in characteristic 0.

In his Part I [Has36c] he considers the case of an elliptic function field $F|K$ with algebraically closed base field. The elements of the Jacobian J are represented by the prime divisors P of $F|K$. (See page 92). In order to prove (8.11) Hasse shows that the primes $P \neq P_\infty$ annihilated by n are contained in the numerator of the differential determinant for the module $\mathcal{L}(P_\infty^n)$, the K-module of multiples of P_∞^{-n}. (Remember: P_∞ denotes the prime of $F|K$ which represents the zero element of the addition of points.) By a detailed study of the prime decomposition of this determinant he gets (8.13).

Today it seems not to be widely known that the theory of higher derivations in characteristic p had been developed by Hasse just for the purpose of his proof of (8.13) in Part I. Hasse reveals to us how he got the idea to use those differential determinants. He says in [Has36c]:

"Here I am employing a method which is well known from the theory of Weierstrass points, namely the use of differential determinants. Thereby I am relying on the theory of higher differentials which is freed from restrictions due to the characteristic, as developed in the foregoing article." (Hasse refers to his paper [Has36a].)

Hasse's foundation of higher differentials was later simplified and generalized by Teichmüller [Tei36], and once more by F.K. Schmidt [Has37c] who developed the theory for arbitrary fields of characteristic p. Moreover F.K. Schmidt used these new differential determinants to develop the theory of Weierstrass points for function fields of genus > 1 in characteristic p [Sch39]. He found the situation quite different from the classical case in characteristic 0. As to the latter he, like Hasse, refers to Hensel–Landsberg [HL02] in characteristic 0.

8.4   The Structure of the Endomorphism Ring

After Hasse had proved the RHp for elliptic fields he was working on function fields of higher genus and left the elliptic case aside. But Deuring continued the work on the elliptic case, which he published in a series of papers between 1940 and 1949. (For biographical facts on Deuring see Sect. 9.2.1.) He determined completely the structure of the endomorphism ring of elliptic function fields. It appears appropriate to report here already about these results although this means a jump in time from 1935 to the 1940s. In between Deuring delivered two other important papers towards the RHp for function fields of genus $g > 1$, these will be discussed in the next chapter.

As above, $F|K$ denotes an elliptic function field with base field K algebraically closed of characteristic p, and let M denote the ring of endomorphisms of its Jacobian J. Hasse had proved that M has no zero divisors and is of characteristic 0. Every element $\mu \in$ M satisfies a monic quadratic equation over $\mathbb{Z}$ with constant term $\mathcal{N}(\mu) > 0$. (See Sect. 7.4.3.) Consequently, its field of quotients Quot(M) is of one of the three following types, and M consists of integers in this field:

I.   Quot(M) $= \mathbb{Q}$.
II. Quot(M) is an imaginary quadratic number field over $\mathbb{Q}$.
III. Quot(M) is a definite quaternion algebra over $\mathbb{Q}$.

Whereas the appearance of types I and II was familiar from the theory of complex multiplication in characteristic 0 already, type III appears in characteristic $p > 0$ only. Hasse comments in [Has36b] on this as follows:

"The example $p = 3$, $y^2 = x^3 - 2x - 1$ shows that type III does in fact occur."

But Hasse didn't say more. The available evidence points to the conclusion that Hasse at that time did not know much more about type III, besides this example and maybe some others.

8.4.1   The Supersingular Case

Deuring in his 1941 paper on endomorphism rings called the fields with endomorphism ring of type III *"supersingular"*. His motivation for this terminology was as follows:

In the classical case, when the base field $K = \mathbb{C}$ is the field of complex numbers, any elliptic function field $F|\mathbb{C}$ can be generated by an equation of Weierstrass form (7.5). Let

$$j = 12^3 \frac{g_2^3}{\Delta}, \qquad\qquad \Delta = g_2^3 - 27g_3^2 \qquad\qquad (8.19)$$

be the invariant of $F|\mathbb{C}$. It is well known that $F|\mathbb{C}$ is uniquely determined (up to isomorphisms) by j, and that every $j \in \mathbb{C}$ is the invariant of some elliptic field $F|\mathbb{C}$. Now, in the classical case the invariant j was called *"singular"* if the endomorphism ring of the corresponding elliptic field is an order in an imaginary quadratic field, i.e., of type II. From classical complex multiplication it was known that these singular invariants j are algebraic numbers, and they are abelian over the corresponding imaginary quadratic field. Thus they are very special complex numbers. Classically the terminology "singular" expresses the fact that these numbers are quite special, in contrast to the "general" case in which j may be a "general" complex number and the ring of endomorphisms is $\mathbb{Z}$.

In characteristic $p > 0$, Deuring used essentially the same terminology. He called an elliptic field $F|K$ (or its invariant j) "singular" if the endomorphism ring is of type II. But in characteristic $p > 0$ the endomorphism ring may be even larger, namely non-commutative of type III. These fields, or their invariants, are somewhat more singular than the others, and so Deuring called them "supersingular".

This was the motivation for Deuring to introduce the word "supersingular".

As to the invariant j of an elliptic function field of prime characteristic p, it is defined for $p > 3$ by the same formula (8.19) as above. Note that for $p > 3$ every elliptic function field $F|K$ admits a Weierstrass normal form (7.5) (if K is algebraically closed). I will discuss below the definition of the invariant in characteristics $p = 3$ and $p = 2$.

In his paper [Deu40] Deuring shows that if j is supersingular then the p-torsion of the Jacobian vanishes. We know that this is equivalent to the vanishing of the Hasse invariant A, see (8.13). But it was not yet clear whether, conversely, $A = 0$ would imply j to be supersingular. This is indeed the case, and it is one of various results which Deuring proved in his long 1941 paper on endomorphism rings for elliptic fields [Deu41a].

Hasse was well aware of the fact that the abstract definition of A as given in Sect. 8.1, would be of no use if one could not compute his invariant A directly. In principle, of course, this can be done by computing the coefficients of the expansion (8.1). For this purpose, one has to start with an explicit expression of the integer differential ω. Suppose for the moment that $p > 3$ so that F can be generated in the form $F = K(x, y)$ with x, y related by an equation in Weierstrass normal form (7.5). Then the integer differential ω can be chosen as $\omega = \frac{dx}{y}$. Expanding this at the point at infinity with respect to the prime element $t = \frac{-2x}{y}$ we get the coefficients c_i in (8.1) as functions of the coefficients g_2, g_3 in (7.5). Now, Hasse in his 1934 paper on unramified cyclic extensions [Has34b] had found that A can be put into the following form, where Δ denotes the discriminant and j the invariant (8.19):

If $p > 3$ the Hasse invariant A defined by (8.3) is of the form:

$$
A = \begin{cases}
\Delta^{\frac{p-1}{12}} P(j) & \text{for} & p \equiv 1 \ \mathrm{mod}\ 12 \\
g_2 \Delta^{\frac{p-5}{12}} P(j) & \text{for} & p \equiv 5 \ \mathrm{mod}\ 12 \\
g_3 \Delta^{\frac{p-7}{12}} P(j) & \text{for} & p \equiv 7 \ \mathrm{mod}\ 12 \\
g_2 g_3 \Delta^{\frac{p-11}{12}} P(j) & \text{for} & p \equiv 11 \ \mathrm{mod}\ 12
\end{cases}
\tag{8.20}
$$

where $P(X)$ is a polynomial, depending on p only, with coefficients in the prime field $\mathbb{F}_p$, of degree at most equal to the exponent of Δ in the formula.

But Hasse adds:

"I suspect that $P(X)$ is always of this degree. But the preceding arguments do not even show whether $P(X)$ is never identical zero."

It is with this question that Deuring begins in his 1941 paper on endomorphism rings [Deu41a]. There he proves the following

Theorem

1. *The relation $A = 0$ is not only necessary but also sufficient for j to be supersingular.*
2. *As conjectured by Hasse, the polynomial $P(X)$ has precisely the degree which is given by the exponent of Δ in the formulas (8.20), and its roots are mutually different. Consequently, the number of supersingular*

*invariants j equals that degree, plus one additional invariant in the cases
$p \equiv 5,7$ mod 12 and two additional invariants if $p \equiv 11$ mod 12. These
additional invariants correspond to the cases $g_2 = 0$ (hence $j = 0$) and
$g_3 \doteq 0$ (hence $j = 12^3$) respectively.*

3. *Supersingular invariants j satisfy $j^{p^2} = j$, hence they are contained in
   $\mathbb{F}_{p^2}$, the quadratic extension of the prime field $\mathbb{F}_p$. (Ogg has later proved
   that there are precisely 15 primes for which all supersingular invariants
   are contained in $\mathbb{F}_p$ already: $2 \leq p \leq 31$ and $p \in \{41, 47, 59, 71\}$. See
   [Ogg75] and also [Mor07].)*
4. *If j is supersingular then the corresponding endomorphism ring M is
   isomorphic to a maximal order in the quaternion division algebra $\mathbb{H}_{\infty, p}$
   which is ramified at ∞ and p only. Conversely, every maximal order
   in $\mathbb{H}_{\infty, p}$ appears as the endomorphism ring M for some supersingular
   invariant. If the prime ideal of p in M is principal then there is exactly one
   supersingular invariant j belonging to M, and j is contained in the prime
   field $\mathbb{F}_p$. If not, then there are exactly two such supersingular invariants j,
   they are contained in $\mathbb{F}_{p^2}$ and they are conjugate to each other.*
5. *The number of supersingular invariants equals the class number of $\mathbb{H}_{\infty, p}$.*

In addition, Deuring writes down an explicit formula for the polynomial $P(j)$
which, he says, is useful to compute the values of the supersingular invariants for
small p.[1] In fact, his paper contains a list of all supersingular invariants for primes
$p < 100$. (In the paper [BM04] the authors state that they have checked the entries
in Deuring's table and found only two errors, for $p = 73$ and 97.)

These results on the supersingular case are very precise and complete. Although
in our discussion we had assumed $p > 3$, it turns out that the theorem holds also in
characteristic $p = 3$ and $p = 2$, except of course its second section which refers to
the Weierstrass normal form (7.5). The definition of the invariant j for $p = 2$ and
$p = 3$ is as follows:

Classically, besides the Weierstrass normal form there is another normal form,
called Legendre's, which is as follows:

$$y^2 = x(x - 1)(x - \lambda) \qquad \text{with} \qquad \lambda \neq 0, 1. \tag{8.21}$$

Then

$$j = 2^8 \frac{(1 - \lambda(1 - \lambda))^3}{\lambda^2(1 - \lambda)^2} \tag{8.22}$$

This works also in all prime characteristics $p \geq 3$.

---

[1]The investigation of those polynomials for the supersingular invariants has produced a number
of highly interesting papers, some of them connecting to the theory of modular forms. See, e.g.,
the list of references in [Mor06]. I would like to thank Patrick Morton for pointing out to me that
those papers arose from the interest generated by Deuring's paper. In particular the question of
determining directly the number of roots of $P(j)$ was solved, which Deuring could solve only
indirectly by means of Eichler's class number formula for quaternions [Eic37].

Referring to the Legendre normal form, Deuring gives a formula for the computation of the Hasse invariant A by means of the parameter λ, namely:

$$A = (-1)^{\frac{p-1}{2}} \sum_{0 \le i \le \frac{p-1}{2}} \binom{\frac{p-1}{2}}{i}^2 \lambda^i . \tag{8.23}$$

For $p = 3$ this reduces to $A = -(1 + \lambda)$ and we see that this vanishes for $\lambda = -1$ only which gives $j = 0$ in characteristic 3. Thus in characteristic 3 there is only one supersingular invariant, $j = 0$. This belongs to the example which Hasse had found in characteristic 3 and which we have mentioned on page 118.

In characteristic 3 it is sometimes easier to work without Legendre's normal form, and use instead two possible normal forms in the style of Weierstrass:

$$y^2 = \begin{cases} x^3 - x^2 - g_3 & \text{if } j = g_3^{-1} \\ x^3 - x & \text{if } j = 0 . \end{cases} \tag{8.24}$$

In characteristic 2 the situation is quite different. Before Deuring, there did not exist a definition of an absolute invariant j of an elliptic function field F of characteristic 2. In this case he obtained the normal forms

$$y^2 - y = \begin{cases} ax^2 + \dfrac{1}{ax} & \text{with } a \ne 0 \\ x^3 \end{cases} \tag{8.25}$$

and Deuring defines $j = a$ or $j = 0$ respectively in these cases.

8.4.2   Singular Invariants

In his 1941 paper Deuring also treats *singular invariants* in characteristic p, i.e., those invariants where the endomorphism ring is an order in an imaginary quadratic field. Perhaps, from a systematic point of view I should have reported about the singular case first before discussing the supersingular case. But I have decided to start with the supersingular case because those results are particularly interesting in view of the fact that non-commutative endomorphism rings do not appear in the classical case and hence represented new discoveries in the 1930s.

Deuring's results on singular invariants are as follows. Recall that $F|K$ is assumed to be an elliptic function field of characteristic $p > 0$ with the base field K algebraically closed. Again, j denotes the absolute invariant of $F|K$ and M its endomorphism ring.

Theorem

1. *j is singular if and only if it is absolutely algebraic* (i.e., algebraic over the prime field $\mathbb{F}_p$) *and* $A \neq 0$.

2. *If j is singular then the corresponding endomorphism ring* M *is isomorphic to an order in an imaginary quadratic field* K $=$ Quot(M) *with the following specifications: p splits in* K *into two different prime ideal factors, and the conductor of* M *is prime to p. Conversely, every order* M *of an imaginary quadratic field* K *with these properties appears as the endomorphism ring belonging to some singular invariant j in characteristic p. The number of singular invariants belonging to* M *equals the class number of* M.

3. *Let* $\mathfrak{p}$ *denote one of the two prime ideal factors of p in* M. *If f is the order of* $\mathfrak{p}$ *in the class group of* M *(i.e., f is the first exponent such that* $\mathfrak{p}^f$ *is a principal ideal) then j is of degree f over the prime field, i.e.,* $\mathbb{F}_p(j) = \mathbb{F}_{p^f}$. *The f conjugates of j are also singular invariants belonging to the same endomorphism ring* M.

Recall that the conductor of M is defined to be the smallest positive number $m \in \mathbb{Z}$ such that every integer $\alpha \in$ K with $\alpha \equiv 1 \bmod m$ is contained in M. It is well known that M is uniquely determined by m, and consists of all integers $\alpha \in$ K which are congruent modulo m to some $a \in \mathbb{Z}$. The (fractional) ideals of M which are relatively prime to the conductor form a group. The class group (modulo principal ideals) is finite, and the number of its elements is the *class number* of M.

Whereas in characteristic 0 *every* order in an imaginary quadratic field is the endomorphism ring of some elliptic function field, Deuring had discovered that in characteristic $p > 0$ this is not so. The restriction concerns the behavior of the characteristic p as an ideal in M. This is the result of Deuring's detailed study of the ℓ-adic representation of M for every prime number ℓ including $\ell = p$ when the representation is 1-dimensional.

The remaining invariants in characteristic $p > 0$, i.e., those which are neither singular nor supersingular, are the transcendental ones. They are precisely the invariants with endomorphism ring M $= \mathbb{Z}$.

8.4.3 Elliptic Subfields

I will not give here a detailed report on Deuring's proofs of the above cited two theorems. These proofs, although not particularly difficult, are somewhat roundabout and not straightforward. But I wish to present here the main ideas of Deuring because they are quite remarkable. Moreover they are essential tools for Deuring's further investigations concerning the algebraic foundation of classical complex multiplication (starting 1949 in the *Hamburger Abhandlungen* [Deu49]).

One of those main ideas is to study the elliptic subfields of the given elliptic field $F|K$.

Recall that for $0 \neq \mu \in \mathsf{M}$ we have denoted by $F\mu$ the image of F under the normalized meromorphism μ. (See page 94.) Now, if $0 \neq \mathfrak{a} \subset \mathsf{M}$ denotes any left ideal of M, let $F\mathfrak{a}$ denote the field theoretic compositum of all $F\mu$ with $\mu \in \mathfrak{a}$. This is an elliptic subfield of F. Deuring showed:

The field degree $[F : F\mathfrak{a}]$ equals the norm $\mathcal{N}(\mathfrak{a})$ of the ideal $\mathfrak{a}$. The Galois group of $F|F\mathfrak{a}$ consists of all translation automorphisms τ_P with P in the kernel $J_\mathfrak{a} \subset J$ of the ideal $\mathfrak{a}$. The endomorphism ring of $F\mathfrak{a}$ is the right order of $\mathfrak{a}$ in M.

Since F may be inseparable over $F\mathfrak{a}$, the Galois group is to be interpreted as the Galois group of the maximal separable subextension. The right order of $\mathfrak{a}$ consists of all elements ρ in the quotient field of M for which $\mathfrak{a}\rho \subset \mathfrak{a}$.

These theorems exhibit a close relationship between the subfield structure of F and the ideal structure of its endomorphism ring M. Moreover:

If F is supersingular then every elliptic subfield of F is of the form $F\mathfrak{a}$ for some left ideal $\mathfrak{a} \subset \mathsf{M}$.

This turns out to be the main reason for the validity of the theorem in Sect. 8.4.1 in the supersingular case (page 119).

In the singular case there are more elliptic subfields. Note that every elliptic subfield $F' \subset F$ defines (by means of the norm $N_{F|F'}$) an endomorphism from the Jacobian $J(F)$ to the Jacobian $J(F')$. The field F' is uniquely determined by the kernel of this endomorphism, together with the degree of inseparability of $F|F'$. The kernel is a finite subgroup of $J(F)$, and the translation automorphisms τ_P with P in this kernel constitute the Galois group of $F|F'$. Conversely, given any finite subgroup of $J(F)$ the translation automorphisms belonging to this subgroup determine a subfield F' of F consisting of the elements fixed by those automorphisms; this is an elliptic subfield of F, as well as any purely inseparable subfield of F'.

Based on these facts, combined with a detailed study of the local representations of M (including the p-adic representation for the characteristic p), Deuring shows:

Suppose F is singular, i.e., M is an order in an imaginary quadratic field K. If $F' \subset F$ then the endomorphism ring M' of F' is an order in the same field K. Conversely, any order M' of K is the endomorphism ring of some elliptic subfield $F' \subset F$ – provided that M' satisfies the condition set forth in the theorem of section 8.4.2, i.e., the conductor of M' is prime to the characteristic p.

The above result is used by Deuring for the existence proof of singular elliptic fields F in characteristic p with prescribed endomorphism rings in a given imaginary quadratic field K. For, by the above result he needs only to construct one elliptic field F with an endomorphism ring being contained in K; then among its elliptic subfields there will appear one with the prescribed order in K. Of course,

K has to satisfy the condition set forth in the theorem of Sect. 8.4.2, namely that
p splits in K in two different prime ideals. And the prescribed order has to have
conductor prime to p.

8.4.4 Good Reduction

In characteristic 0 it is well known from analytic uniformization that every order M
in an imaginary quadratic field is the endomorphism ring of some suitable elliptic
function field. One has to view M as a 2-dimensional lattice in $\mathbb{C}$ and then take F to
be generated by that Weierstrass function $\wp(u)$ which has period lattice M, and its
derivative $\wp'(u)$.

In characteristic p one has to assume that M satisfies the specifications set forth
in the theorem stated in Sect. 8.4.2. But there is no direct way yet of proving
the existence of an elliptic field F with a given such M as its endomorphism
ring—except to construct F as a good reduction of a suitable function field in
characteristic 0. In order to be able to do this, Deuring had to establish the necessary
tools from the theory of good reduction.

More precisely, he had to *develop* the theory of good reduction since until that
time no systematic way of reducing curves was known. It is true that Hasse in his
first proof already used the idea of lifting an elliptic curve in characteristic p suitably
to characteristic 0 and then studying the behavior of the lifted curve by reducing it
again. But he had no general theory of reduction at his disposal; therefore he had
to check directly every detail in the reduction process. Since he relied on explicit
computations with generating equations, this resulted in several restrictions which
he had to impose, e.g., the characteristic should be $p > 3$, and the invariant j
of the elliptic curve should have odd degree over $\mathbb{F}_p$. (See Sect. 7.3.) But also in
Hasse's second proof which works solely in characteristic p, he had to use several
constructions which today we would subsume under the theory of good reduction.
(See Sect. 7.4.2.)

Now, Deuring wished to systematize all those arguments by a general theory of
good reduction; he did it in his 1942 paper [Deu42]. Although that paper appeared 1
year later than his 1941 paper [Deu41a] on endomorphism rings, it was completed at
the same time, and Deuring relies on it in his 1941 paper. What are the main results
which Deuring had achieved?

Deuring's theory of good reduction refers to the following situation: Given an
algebraic function field $F|K$ (or curve) whose base field K is equipped with a prime
$\mathfrak{p}$, i.e., valuation, or place. The characteristic of K may be 0 or $p > 0$. Deuring
assumed that the valuation is discrete but a straightforward check shows that this
assumption is not really necessary. $\mathfrak{p}$ can be any valuation, or place, in the general
sense of Krull. Accordingly we may keep our general assumption that the base field
K is algebraically closed, whereas Deuring worked with discrete valuations only
and therefore had often to perform a finite extension of the base field in order to

ensure the validity of his argument. The residue map (place) of K belonging to $\mathfrak{p}$ is denoted by $z \mapsto z\mathfrak{p}$. We also write $\bar{z}$ instead of $z\mathfrak{p}$.

Suppose that $\mathfrak{p}$ can be extended to a place $\mathfrak{P}$ of F with the following properties:

1. The residue field $\overline{F} = F\mathfrak{P}$ is an algebraic function field with base field $\overline{K} = K\mathfrak{p}$.
2. There exists a separating element $x \in F$ such that $\bar{x} = x\mathfrak{P}$ is transcendental over $\overline{K}$ and $[\overline{F} : \overline{K}(\bar{x})] = [F : K(x)]$.
3. The genus of $\overline{F}$ equals the genus of F.

(Recall that we have assumed K and hence $\overline{K}$ to be algebraically closed. Much of Deuring's theory remains true without this assumption; then one has to add the condition that $F|K$ and $\overline{F}|\overline{K}$ are conservative, i.e., their genus should be preserved under extensions of the base field.)

In the above situation $\overline{F}|\overline{K}$ is called a "good reduction" of $F|K$ at $\mathfrak{p}$. Actually, Deuring did not use the terminology of "good reduction" which was introduced later. Deuring spoke of a "regular reduction", and $\mathfrak{P}$ was called a "regular extension" of $\mathfrak{p}$ to F. For a given $F|K$, almost all primes $\mathfrak{p}$ of K (i.e., all but the poles of finitely many elements) admit a regular extension $\mathfrak{P}$ to F. Deuring in [Deu42] did not yet know that the regular extension $\mathfrak{P}$ of $\mathfrak{p}$, if it exists, is unique if the genus of $F|K$ is > 0. For genus $g = 1$ he proved it later in [Deu55], and for arbitrary $g > 0$ this was shown by Lamprecht [Lam57]. In the following let us assume that $g > 0$.

If $\overline{F}|\overline{K}$ is a good reduction of $F|K$ in the above sense then this leads, according to Deuring, to a "reduction map" of the divisor group $\mathrm{Div}(F|K)$ to the divisor group $\mathrm{Div}(\overline{F}|\overline{K})$ such that the relations between divisors, elements and divisor classes are preserved. In particular this means that integer divisors are mapped to integer divisors, the image of a divisor has the same degree as the divisor itself, principal divisors are mapped to principal divisors, etc. More precisely, if A is a principal divisor in $\mathrm{Div}(F|K)$, say $A = (z)$ with $0 \neq z \in F$ then z can be normalized by a factor from K such that its residue $\bar{z} = z\mathfrak{P} \neq 0, \infty$, and then the image $\overline{A}$ of A equals the principal divisor $(\bar{z})$.

(Deuring's theory of good reduction was later generalized by Shimura [Shi55], in the framework of algebraic geometry to varieties of arbitrary dimension.)

This being said, Deuring in his 1941 paper on endomorphism rings shows:

Suppose $F|K$ to be elliptic. Then:

(1) *$F|K$ admits good reduction at $\mathfrak{p}$ if and only if its invariant j is $\mathfrak{p}$-integer, i.e., $j\mathfrak{p} \neq \infty$. If this is the case then the absolute invariant of $\overline{F}|\overline{K}$ is the image $\bar{j} = j\mathfrak{p} \in \overline{K}$.*
(2) *If $F|K$ admits good reduction at $\mathfrak{p}$ then there is a natural isomorphism $\mu \mapsto \overline{\mu}$ of the endomorphism ring M of F into the endomorphism ring $\overline{\mathsf{M}}$ of $\overline{F}$ such that $\overline{\mu P} = \overline{\mu}\,\overline{P}$ for any point P of F and its reduction $\overline{P}$ of $\overline{F}$.*

M may be identified with its image in $\overline{\mathsf{M}}$ so that

$$\mathsf{M} \subset \overline{\mathsf{M}},$$

and then the formula in (2) appears as

$$\overline{\mu P} = \mu \overline{P}.$$

$\overline{M}$ may be larger than M. If F is singular then $\overline{F}$ is either singular or supersingular. If both F and $\overline{F}$ are singular then M and $\overline{M}$ have the same quotient field. If in addition M is a *maximal* order in its quotient field then $\overline{M} = M$.

The following theorem can be used to lift an elliptic function field from characteristic p to characteristic 0.

Theorem *As above, suppose K equipped with a place $\mathfrak{p}$, and let $\overline{K} = K\mathfrak{p}$ be its residue field. Let $\overline{F}|\overline{K}$ be a given elliptic function field, and μ one of its endomorphisms. Then there exists an elliptic function field $F|K$ admitting $\overline{F}|\overline{K}$ as a good reduction modulo $\mathfrak{p}$ such that its endomorphism ring M contains μ.*

In other words: The elliptic function field $\overline{F}|\overline{K}$, equipped with a given endomorphism, can be lifted from the residue field $\overline{K}$ to K.

This is an important result. It explains and systematizes Hasse's procedure in his first proof of the RHp (see Sect. 7.3). Consider the situation of the RHp, i.e., an elliptic curve defined over a finite field with $q = p^r$ elements (see Sect. 7.4.4). Hasse, in his first proof not yet being aware of the notion of Frobenius operator on the Jacobian, succeeded somehow to lift the elliptic function field to characteristic 0 such that after lifting, q splits in the endomorphism ring into two factors, one of them called π. Using class field theory and reduction modulo a prime divisor of p, Hasse verified that this π has the properties which we now use to define the Frobenius operator.

8.5   Class Field Theory and Complex Multiplication

Deuring's proof of the above theorem contains an interesting detail. In order to lift an elliptic function field with a given endomorphism from characteristic $p > 0$ to characteristic 0 one has to know beforehand that there exists a singular curve (or invariant) with an endomorphism ring belonging to a given imaginary quadratic field. Classically this is easy by using the Weierstrass function $\wp(z|\mathfrak{w})$ with $\mathfrak{w}$ in the given imaginary quadratic field, as explained on page 86. But this argument uses the theory of analytic functions.

Deuring's achievement consists of a purely algebraic argument for this purpose. That is, he does not use analysis. Once this is achieved, the classical theory of generating class fields over imaginary quadratic number fields can be completely reformulated in algebraic terms. Instead of generating class fields by "singular" values of analytic functions one can now find the generators essentially as points on the Jacobian of the corresponding algebraically constructed curve.

In other words: Deuring had freed the class field theory of complex multiplication from analysis and gave a purely algebraic treatment. His final proof is published in the year 1949 in the "Hamburger Abhandlungen" [Deu49]. But already in 1939 he had given a colloquium talk in Hamburg about this topic. The publication had been delayed in war time.

I do not wish to go into more detail of Deuring's work in this direction because this would lead us too far from the RHp. But I would like to mention on this opportunity the following citation.

For, already in 1936 Deuring had been in the possession of this theory, and Hasse mentioned this in his letter to Weil dated 12 July 1936 (see Sect. 9.4). The last paragraph of that letter reads as follows:

"By the way, Deuring has still another application of his theory. He wants to solve Hilbert's problem of constructing class fields of algebraic number fields. In the elliptic case he has already been able ... to do this. He even manages this with purely algebraic methods, without using the theory of elliptic functions as functions of a complex variable, i.e., solely using the arithmetic theory of algebraic functions. This is very appealing to me. I have nothing against using the beautiful theory of abelian functions of complex analysis, but I do prefer a 'methodenrein' justification of purely arithmetical facts ... "

REMARK For arbitrary number fields this was wishful thinking. In fact, Deuring considered imaginary quadratic number fields and their class fields only. The algebraic theory of class fields over so-called CM-fields, related to abelian varieties, was established much later by Shimura, Weil, and others.

I have cited the above text since here appears the word "*methodenrein*". I have not found an English translation of this word. It expresses what Hasse had said also on other occasions, that he considers a mathematical theorem and its proof as a unit and hence they should fit together in a natural way, and should be adequate to the problem. Of course, it cannot be absolutely defined which methods are "adequate" in this sense; this depends very much on the mathematical background and the psychology of the respective person. For example, Hasse preferred the theory of algebraic function fields whereas Weil tended to algebraic geometry in this connection.

But it should not be overlooked that Hasse did not insist on "methodenrein" proofs. In fact, he says explicitly that he has nothing against using abelian functions of complex analysis in this context. In fact, he had several times expressed his view that classical complex multiplication is a beautiful example of the cooperative interrelation of the three main mathematical fields, namely Analysis, Algebra and Arithmetics (the three Gaussian "A"s).

By the way, Hasse himself had foreseen that through his algebraic treatment of complex multiplication in characteristic p it would eventually be possible to algebraize the whole classical body of analytically based complex multiplication in characteristic 0. In his lecture 1934 in Hamburg (mentioned in Sect. 7.4) Hasse had said:

> "*Starting from the analytic theory of elliptic functions I have already last year ... sketched a proof of the Riemann hypothesis based on the class field decomposition law of extension fields of imaginary quadratic number fields, namely those extensions which are generated by the division values of the elliptic functions. From this viewpoint it is understandable that an algebraically based proof of the Riemann hypothesis leads, in the other direction, to a proof of the said class field decomposition law ... Moreover, this kind of reasoning appears much more natural.*"

This was just a vision, for at the time of writing this text Hasse had not yet worked out all details, as we have seen above. Finally it turned out to be Deuring who completed this project of Hasse's.

Summary

While Hasse worked on the RHp for elliptic function fields he and his team discovered several properties of elliptic function fields in characteristic $p > 0$, which were new and important although not all of them were necessary for the proof of RHp. The first and most important is the so-called Hasse-invariant of an elliptic function field $F|K$ which is usually denoted by A. This is an element in the base field K. It has no analogue in characteristic 0. Actually A is not really an invariant of the function field, but it is determined up to a substitution $A \to c^{-(p-1)}A$ with $0 \neq c \in K$. If K is algebraically closed then the alternative $A = 0$ or $A \neq 0$ is an invariant of the function field. A is defined as the first obstacle when trying to integrate the integer differential ω of the elliptic function field (ω is unique up to a nonzero factor of the base field).

This invariant A of an elliptic function field $F|K$ controls, among others, the existence of unramified extensions of degree p, the structure of the p-subgroups of the Jacobian, and the structure of the endomorphism ring of the Jacobian. For instance, $A = 0$ signifies that the endomorphism ring is non-commutative— a phenomenon which was not expected and aroused some curiosity. In fact, if $A = 0$ then the endomorphism ring is isomorphic to a maximal order of the quaternion algebra which is ramified at p and ∞ only. Deuring has called such elliptic field $F|K$ "supersingular". The invariant A can be computed with the help of an explicitly given polynomial $P(j)$ where j is the absolute invariant of the elliptic field $F|K$. (If $p > 3$ then this invariant j can be defined and computed by the coefficients of the Weierstrass normal form in the usual way, but for $p = 2$ and $p = 3$ this does not work and j has to be defined differently.) The degree of the polynomial $P(j)$ depends on p. For given p there are only finitely many j for which $A = 0$, and they can be explicitly computed.

With the help of these and more results Deuring was able to give a simple, structural proof of Hasse's lifting theorem which had been used in Hasse's first but unpublished proof of the RHp for elliptic fields. This gives a purely algebraic proof also of the theory of classical complex multiplication in characteristic 0. Hasse had foreseen this possibility already in his Hamburg lectures in January 1934.

Chapter 9
Towards Higher Genus

9.1 Preliminaries

As I have told in Sect. 6.2 Hasse originally was interested in the estimate of solutions of diophantine congruences. It was Artin who in November of 1932, when Hasse was visiting Hamburg, told him that the estimating problem was pointing to the RHp. It appears that at that time Hasse did not yet believe in the general validity of the RHp for *all* function fields with higher genus $g > 1$. (See page 65.)

But shortly thereafter Hasse had changed his mind. On 6. March 1933 when he reported to Mordell that he had found his (first) proof of the RHp for elliptic fields (see page 82) he closed his letter with the words:

> "*Obviously the general congruence $f(x, y)$ may be treated the same way, "only" with the "slight" generalization of the elliptic functions into abelian functions quite analogous to Siegel! Do it!*".

Here Hasse refers to Siegel's work on diophantine approximations which he had studied 1 year earlier together with Mordell. I have mentioned this already on page 88.

But Mordell threw the ball back to Hasse. In his reply (dated 9. March 1933) he wrote:

> "*I think the results for $y^2 \equiv f_n(x)$ etc. should follow without infinite difficulty, but the zeta fn. theory will not be so simple now. Obviously you are now the one to try it.*"

A little later, on 14. April 1933 in Hasse's preliminary announcement to the Göttingen Society of Science he again pointed to the possible generalization to

© Springer Nature Switzerland AG 2018

P. Roquette, *The Riemann Hypothesis in Characteristic* p
in Historical Perspective, Lecture Notes in Mathematics 2222,
https://doi.org/10.1007/978-3-319-99067-5_9

higher genus [Has33a]. He said that the proof in the general case could probably be achieved through

"...*uniformization of binary diophantine congruences by general abelian functions.*"

This refers to his first proof in the elliptic case where he had established some kind of uniformization of elliptic congruences by means of the complex Weierstrass ℘-function. However, Hasse continues,

"...*it would first be necessary to develop complex multiplication for general abelian functions if not the arithmetic-algebraical methods of A. Weil would be already sufficient.*"

Here, Hasse refers to Weil's thesis [Wei29] which contained the so-called Mordell-Weil Theorem about the finite basis of the rational points of a Jacobian variety over a number field.

It appears that Hasse's ideas were still somewhat vague at that time, without any definite plan how to attack those problems. In a letter from Hasse in Marburg to Davenport in Göttingen dated 24 July 1933[1] we read that "*Weil came over for a day*". There is no doubt that on this occasion Hasse and André Weil had discussed problems of the RHp for higher genus. I do not know whether Weil's visit was the answer of a letter from Hasse, or perhaps he had heard of Hasse's ideas and came over to obtain more information.

Observe that this happened in the summer of 1933 when Hasse still worked on his first proof in the elliptic case, using classic complex multiplication over an imaginary quadratic number field. It seems that also for higher genus he had in mind the classic theory of abelian functions based on ϑ-functions.

But next year, when Hasse in February 1934 gave a series of lectures in Hamburg about his second proof in the elliptic case, he mentioned in the introduction the possibility of an *algebraic* theory of abelian function fields, valid also in characteristic p. He said:

"*In the following I am developing the essential features of the theory for the simplest non-trivial case, i.e., the case of elliptic function fields. This seems similar to Hilbert's theory of relative quadratic fields which he developed 36 years ago as the simplest case of the relative abelian fields which he sketched later – but he always had this generalization in mind.*"

Hasse refers to Hilbert's paper from the year 1898 [Hil98].

Well, without comparing Hasse with Hilbert, we shall see below that the solution of the problem would finally *not* be given by Hasse, like in Hilbert's case where the

---

[1]In the summer term of 1933 Davenport stayed in Göttingen with a stipend he had got from his College. Thus he became a witness of the dissolution of the Göttingen mathematical scene due to the disastrous policy of the Nazi government. On weekends he often went to Marburg, which is not far from Göttingen, in order to meet Hasse. But sometimes there were letters exchanged.

final building of class field theory was not given by Hilbert. In Hilbert's case, class field theory was completed by Takagi and Artin, while in Hasse's case we shall see that the proof of the RHp for arbitrary genus was finally completed by A. Weil. But like Hilbert in the case of class field theory, Hasse in the case of the RHp had already envisaged the essential tools which would lead to the goal.

During Hasse's stay in Hamburg he discussed with Artin the possible generalization to function fields of genus > 1. He reported on this to Davenport in his letter dated 12 February 1934 as follows:

"From what Artin and I found when considering the possibilities of generalization to higher genus, it will be only a matter of patience to do this. The general line is fully obvious now. The addition theorem is generalizable in a purely algebraic form. If $f(x, y) = 0$ has genus g, then for each two sets of g solutions

$$(x_{11}, y_{11}), \ldots, (x_{1g}, y_{1g})$$
$$(x_{21}, y_{21}), \ldots, (x_{2g}, y_{2g})$$

there exists a third such set (uniquely determined, except the arbitrary arrangement)

$$(x_{31}, y_{31}), \ldots, (x_{3g}, y_{3g})$$

with all the algebraic properties of an "addition" of the two sets. The number of automorphisms of the field of all symmetrical functions of g independent solutions

$$(x_1, y_1), \ldots, (x_g, y_g)$$

is finite.[2] This gives the fact that the abstract operation π (defined as p^{th} power) is algebraic of degree $2g$, and that the field of π as an algebraic number contains only $g - 1$ independent units, i.e., is totally–imaginary. I am going to carry through all the details without bothering about any more special cases now."

The "special cases" which Hasse has in mind seem to be the generalized Fermat fields and the Davenport–Hasse fields (see Sect. 6.3).

Here, Hasse expressed clearly his plan what he was going to do in the case of higher genus, following Artin's suggestion. It was necessary to give an algebraic construction of the Jacobian function field of the curve $f(x, y) = 0$, defined by symmetrization of g algebraically independent generic points of the curve. And

---

[2] *It appears that Hasse had those automorphisms in mind which respect the "addition" of two point sets. He knew already from the elliptic case that there are more field automorphisms, namely the translations. See page 95.*

then to study the action of the Frobenius operator π as an endomorphism of the Jacobian variety, with the aim of proving that the algebra generated by π is behaving somewhat like being totally imaginary.

But Hasse's words do not give any evidence that he was aware of the various details of the work ahead. It seems that he had discussed with André Weil his new ideas, i.e., working directly in characteristic p instead of reduction from characteristic 0. I do not know whether and where the two met in the summer of 1934, but the first sentences of Weil's letter, dated 18. June 1934, may be interpreted that such meeting had taken place:

> *"I have again thought about your problem. According to my experiences on this matter it cannot be expected to arrange feasible computations without the theta function. Using homogeneous functions, which we had discussed, is useful to avoid the places of indetermination but yet it is not a tool for computations. In my opinion, you are left with operating with the theta function as usual, after having defined it algebraically. For, the theta function is nothing else than a divisor on the Jacobian variety: the divisor $\vartheta(u_1 - a_1, \ldots, u_g - a_g) = 0$ is a $(g - 1)$-dimensional variety; its points correspond to those divisors $\mathfrak{X}$ on the basic curve in question which are not uniquely determined by their divisor class $\left(\frac{\mathfrak{X}}{\mathfrak{A}}\right)$. The algebraic content of the periodicity properties of the theta-function is nothing else than a necessary and sufficient condition for the quotient*
>
> $$\frac{\prod_\mu \vartheta(m_\mu u - a_\mu)}{\prod_\nu \vartheta(n_\nu u - b_\nu)}$$
>
> *to be an abelian function, which is to say that the corresponding divisor on the Jacobian variety is principal. These properties can probably be proved without great effort, as well as the few additional properties which are used in the computations with theta-functions. In this way you will rediscover the only useful computational apparatus which so far exists in this theory."*

We see that Weil saw very clearly the possibilities and necessities for Hasse's project. He appears to have been keenly interested in the further progress since he concluded his letter as follows:

> *"Please do inform me occasionally about the progress of your investigations. You may always use my address in Paris."*

9.2   The Years 1934–1935

The documents which we have cited above indicate that Hasse was planning to investigate the Jacobian and its endomorphism ring in an algebraic setting. These documents are dated in 1934. But I have no indication that Hasse really started to implement his ideas within the near future. What was the reason?

In the years 1934–1935 Hasse was busy writing up his second proof of the RHp in the elliptic case. The final paper in Crelle's Journal consisted of three parts I,II,III and was completed at the end of 1935 when he submitted the manuscript to Crelle's Journal. In November 1935 he reported about it to the Göttingen Society of Science [Has35]. (I have discussed this in Chap. 7.) Parallel to this, Hasse jointly with Davenport studied what today are called generalized Fermat fields and Davenport–Hasse fields (See Sect. 6.3.) In addition, Hasse and his collaborators were able to establish a number of basic results for arbitrary function fields (or curves) over finite fields, including class field theory, L-series, differentials and their residues, higher derivations, p-extensions, and more. Certainly, this amounted to a lot of work and it may be that Hasse first wished to complete this before turning to the projected algebraization of the Jacobian for genus > 1. In the Hasse papers in Göttingen I have not found any indication for work towards the Jacobian from those years.

We also have to take into account the difficulties of political nature which Hasse had to face after May 1934, when he had changed from Marburg to Göttingen. Those years were the first years of the Nazi government in Germany. Due to the expulsion of the Jewish professors and those who appeared to be in opposition to the government, the mathematical scene was severely hit. Hasse had accepted the offer to Göttingen with the intention to rebuild Mathematics there. From Hasse's correspondence in particular with Davenport I infer that he had been quite aware that he was to face political troubles from the Nazi students and colleagues in Göttingen. In his letter to Davenport of 17 February 1934 he wrote:

"I very much hope that I shall have a quiet summer here before I am called upon the battle field. Otherwise I doubt whether I shall be able to think on $f(x, y) = 0$ for higher genus before long."

Hasse's hope for a quiet summer was in vain. Hasse had to start in Göttingen already in the summer semester. There he met strong and hateful opposition, and also faced perpetual difficulties with the governmental agencies. Much of his energy was absorbed by dealing with those problems. The general story is fairly known and need not be repeated here in detail. (See, e.g., [Sch87]. Compare also the revised online version.)

One of the greater issues, in Hasse's opinion, was the appointment of mathematicians to the Göttingen Institute. Originally Hasse had insisted that only the mathematical standing of the applicant should be relevant, and this was assured him by the representative of the government. However Hasse could not carry his point since in reality the rules of the Nazi government required that also the political standing should be taken into consideration. Let us cite from a letter of Hasse to Toeplitz in Bonn dated 18 April 1935:

"What I find highly depressing is the fact that, on the one hand, I am carrying the responsibility, towards the world of mathematics, for the rebuilding of Göttingen as a high-ranking mathematical site, but on the other hand I have almost no decisive influence on the personnel side of staff appointments, due to the existing political rules."

From the correspondence Hasse-Toeplitz of that year we can see that Hasse even contemplated to leave Göttingen and change to Bonn, precisely because of these obstructions to his activities which he mentions in this letter. Toeplitz, although himself being hit by the Nazi regulations, had seen to it that the University of Bonn proposed to offer Hasse a position in Bonn, which means that the name of Hasse appeared first on their "*Berufungsliste*". However the ministry in Berlin did not follow this proposal from Bonn University, perhaps because they wished to keep Hasse in Göttingen as a prominent figure.

Also, the attempts of Hasse to get Nevanlinna and van der Waerden permanently to Göttingen would turn out not to be realizable. (Finally, Nevanlinna came to Göttingen as a visiting professor in 1936–1937.) But the immediate cause for Hasse's complaint in his letter to Toeplitz was connected with his attempt to install Deuring as a *Dozent* (associate professor) in Göttingen.

9.2.1   Deuring

Max Deuring (1907–1984) had been a student of Emmy Noether in Göttingen. He was her "Lieblingsschüler", as she once called him. He was allowed to call her with the familiar "Du", which at that time was quite unusual in Germany since the usual form of addressing in German was the somewhat more formal "Sie", also among students and colleagues. As a student Deuring had spent 1 year in Rome. In 1930, when he was 22 of age, he got his Ph.D. in Göttingen. One year later he obtained a position as assistant professor in Leipzig with van der Waerden. In the academic year 1932/1933 he stayed at Yale University with Øystein Ore. When he returned to Leipzig he wished to apply for a teaching position (*Dozent*) at the university there. In German universities this is a traditional procedure called "*Habilitation*"; the candidate has to present a thesis to the Faculty and to deliver a lecture to the faculty in order to qualify as *Dozent*.

But in the meantime the Nazis had come to power in Germany and Habilitation had become possible only when the candidate appeared to be politically in line with the government. Deuring had the strong support of van der Waerden in Leipzig but nevertheless he was not admitted to apply for Habilitation there since he failed to obtain the approval of the local Nazi authorities. In this situation Hasse who held a high opinion about Deuring, advised him in agreement with van der Waerden, to apply for Habilitation in Göttingen. But this turned out not to be easy. In the letter to Toeplitz of 18 April 1935 (from which I have cited above already) Hasse wrote:

"Deuring is not yet habilitated. Just now he intends to do this here. I shall have a hard fight for him here. For, his quiet scholarly personality is not of the type which one would prefer here. Mathematically he is first class."

Indeed, in the years 1935–1937 the young Deuring had ten publications, each of first class quality, among them the beautiful book on algebras [Deu35] and his work on function fields, in particular the paper on correspondences [Deu37] which will be discussed below.

From the letters exchanged between Hasse and Deuring it can be seen that the date for Deuring's Habilitation lecture in Göttingen had originally been set for 24 April 1935. But this date was repeatedly postponed by the university, for reasons unknown to me. It may well have been that the authorities waited for background information about the political standing of Deuring. Finally the date was set to 11 December 1935. But the result was negative: The faculty committee did *not* admit Deuring as *Dozent* in Göttingen. This was due to the massive opposition by the Nazi colleagues, in particular Tornier and Teichmüller, as well as by the University Rector (President). Hasse was deeply disappointed, and he wrote to van der Waerden in a letter of 16 December 1935:

"Everyone here who had supported him [Deuring] was deeply disappointed about the outcome. Mr. D. has lost a battle for us, so to speak ..."

When Hasse wrote that Deuring had lost a battle "for us" then apparently he interpreted the decision of the faculty committee as opposition against himself and also against van der Waerden since both, in the eyes of the fervent Nazis, were not in line with the official Nazi *Weltanschauung*.

In fact, there exists a document (*Gutachten*), signed by Tornier and dated 2 May 1935, in which Hasse is denounced as not being suitable for the position as chairman of the Göttingen Mathematical Institute, the main reason being his "Jewish decent" and, consequently, his lack of appreciation of the Nazi ideology. (I have found this document in the *Bundesarchiv* in Berlin.) Moreover, Tornier says, Hasse is not able to judge the mathematical skills of young people which he tries to support. Although the name of Deuring is not mentioned it seems obvious, in view of the date of this document, that it was written in connection with the Deuring affair in Göttingen. I do not know to whom this paper was addressed. Evidently Tornier wished Hasse to be removed from the position as chairman with the hope that he, Tornier, would then be installed in this position.

After the refusal of Deuring in Göttingen, Hasse met van der Waerden in order to discuss with him how Deuring could be supported in the future. Van der Waerden wrote on 17 December 1935:

"It will be of great value to me to have an opportunity to speak with you about Deuring's failure, about his future and what we now have to do. I am grateful to you that you are even willing to come to Leipzig for this, and I am at your disposal the whole day of 4. January."

The outcome of their meeting was that for the time being Deuring remained as assistant professor with van der Waerden in Leipzig. But the possibility of a new application of Deuring for a position in Göttingen was also discussed. In a letter of Hasse to Deuring of 11 June 1936 he reported that the ministry of education had assured him (Hasse) there would still remain the possibility for a new application. Perhaps it would be wise to wait with this for, say, one year. But Hasse adds:

"In my opinion it would not be worthwhile to make a new application as long as the present Rector[3] is in office ... Moreover I cannot guarantee that the Fachschaft[4] will not oppose your application as long as Mr. Teichmüller is still here."

But Deuring did not send a new application, although in the next year Teichmüller left Göttingen. Later (1938) Deuring obtained, with recommendation also by Hasse, a position as *Dozent* (associate professor) at the University in Jena where F.K. Schmidt was full professor. After the war Deuring got a professorship in Marburg (1947), Hamburg (1948) and finally Göttingen (1950).

REMARK I have been advised to say a few words about the role of Teichmüller (1913–1943) in relation to Hasse. Teichmüller was known to be a fervent follower of the Nazi ideology. On the other hand, he was also known as a very gifted mathematics student. He participated in the *Arbeitsgemeinschaft* which was led by Witt, an assistant of Hasse. The subjects of the *Arbeitsgemeinschaft* were set

---

[3]President of the University.

[4]The official organization of mathematics students which at that time was dominated by Nazi students.

by Hasse who regularly participated in the sessions. Hasse was mathematically accepted by Teichmüller but not politically. Some sources say that Teichmüller had been a Ph.D. student of Hasse. But this was not the case. Teichmüller had conceived and written his thesis completely by himself, its subject belonged to functional analysis and seems to have been inspired by Rellich's seminar 1933/1934. When T. finally submitted his thesis to Hasse then Hasse did not feel competent himself for this subject and he asked Köthe to referee it. The latter sent his report to Hasse dated 18 February 1935 and recommended T's thesis. I have taken the above information from the article [SS92] where more details about the life and mathematical work of Teichmüller can be found.

9.2.2 More Political Problems

In a letter to Siegel dated 19 December 1935, Hasse admits that in the past year he had not been able to continue his research as planned, and only recently he had returned to his work on Jacobian function fields. This letter was written as a reply to an earlier letter from Siegel who had asked Hasse for news on function fields of genus $g > 1$. In the preceding year Siegel had been in Princeton (USA) and had met Emmy Noether who was still there. (She died on 14 April 1935 in Bryn Mawr.) She had told him that Hasse was working on this topic, but she was not able to give details. Hasse writes to Siegel:

> "Unfortunately, I am still at the beginnings with the abstract function fields of genus $g > 1$. For, I was forced to interrupt my scientific work thoroughly since May 1934, and not until October this year I was able to return to my research."

May 1934 was the time when Hasse had moved to Göttingen. In October 1935 he had finished the first part of his 3-part paper on elliptic function fields and their RHp.

In his letter to Siegel Hasse describes first what he had been doing in the elliptic case. He had obtained essential simplifications of the proof, he wrote, so that it has become much clearer in view of the intended generalization to higher genus. Obviously Hasse refers to his second proof in the elliptic case. And then he talks about his plans for Jacobian function fields. Given two endomorphisms μ, ν and a function φ on the Jacobian, he wrote, one has to be able to determine, by algebraic means, the number of poles and zeros of the difference $\varphi(\mu z) - \varphi(\nu z)$. Hasse had in mind his proof in the elliptic case of what he had called "Norm Addition Formula" which, he said, indeed constituted the key to his proof of the Riemann hypothesis. (See Sect. 7.4.3, in particular formula (7.21) on page 101.) In the general case, the poles and zeros of a function on the Jacobian are to be divisors which, as Hasse knew at least since his discussion with A. Weil, have to be considered as the algebraic equivalent of the ϑ-functions. But his remarks in the letter to Siegel do not give any hint that so far he has made any essential progress towards a suitable algebraic definition and study of these ϑ-divisors beyond the elliptic case.

Even after October 1935 which Hasse had mentioned to Siegel as the time of a new start of his research activities, there lingered a lot of problems of political kind which he had to deal with. I have already mentioned the affair with Deuring's *Habilitation* in December 1935. Early in 1936 Hasse's colleague Tornier in Göttingen, in addition to having recently barred Deuring's appointment, started a new campaign against Hasse. On the outside it was directed against the German Mathematical Society (DMV) which Tornier openly accused of tolerating Jewish mathematicians among its members. But since Hasse was active in the governing board (*Vorstand*) of the society it is evident that again this move was the attempt to disqualify Hasse, in the eyes of the Nazi authorities, as the chairman of the Göttingen Institute.

This time Hasse went to the ministry of education in Berlin and told them categorically that he will resign from Göttingen University if working conditions would not be improved and, in particular, if Tornier would not be stopped in poisoning the mathematical atmosphere in Göttingen. This time, as a consequence of Hasse's move, Tornier had to leave Göttingen and was assigned to the University of Berlin in April 1936.[5]

Let us cite from a letter of Hasse to F.K. Schmidt dated 25 June 1936:

"Two years ago, when I signed my contract , I harbored the hope that the unsettling troubles would slow down soon. But you know that this happened only in mid-April of this year. Until then, as you can imagine, my working power was quite paralyzed. Any time when I started to intensify my work, after a short time there arose new problems which hampered my work."

(Hasse refers to his contract with Springer Verlag about his planned book "Zahlentheorie". F.K. Schmidt at that time was the managing editor of the "yellow series" of mathematical monographs, in succession to Courant who had been forced by the Nazis to emigrate.)

The reference to "mid-April" refers to the removal of Tornier from Göttingen. But this time again, Hasse's statement that the politically based problems in Göttingen had ceased by mid-April 1936 or at least slowed down, was wishful thinking. Even without Tornier poisoning the atmosphere at the mathematical scene in Göttingen, there remained political trouble there.

---

[5]An additional reason for removing Tornier from Göttingen may have been that he permanently produced financial trouble. He had pawned a large part of his future salary and also had asked for and received high sums from the university as advance payment. It appears that he had fallen back to his addiction to drugs (morphine) which Hasse had mentioned in a letter to Fraenkel of 10 July 1927. (At that time Hasse had believed that Tornier had been able to overcome this.) But at Berlin University Tornier did not stop this behavior and in 1938, when he was caught in financial fraud he had to leave the university and also the Nazi party. More precisely, in order to avoid public scandal he was "advised" to leave the party and the university on his own application. He then retreated to a psychiatric clinic in Silesia. These facts are documented in the personal files for Tornier at the archive of Humboldt University in Berlin.

I have mentioned all this in order to point out that during the years 1934/1935 and later Hasse was seriously hampered in his mathematical research by difficulties which originated in the political situation of the time. The question arises *why* Hasse had shouldered the *"responsibility for the rebuilding of Göttingen as a high-ranking mathematical site"*, as he had written to Toeplitz (see page 136). It had been clear from the outset that he would run into heavy difficulties. Hasse had been warned by several of his friends not to trust the words of assurance by the government authorities, and had been advised to remain in Marburg or even to emigrate from Germany. One of those friends was F.K. Schmidt. His letters to Hasse are preserved. It appears F.K. Schmidt had contacted Courant and inquired about whether Hasse would possibly have a chance at some place in the USA. Hasse's friend Davenport had written on 20 February 1934, somewhat doubtfully:

"What I feel about G[öttingen] is that if it is in the power of one man to restore the prestige of G., you are the man. But perhaps it is not within the power of one man."

But on the other side, Hasse had also been encouraged to accept the Göttingen job, for instance by Hermann Weyl who expected precisely what Hasse had written to Toeplitz, namely that he should rebuild Göttingen as a mathematical center. Even Fraenkel from Jerusalem had sent a letter to Hasse wishing success in this endeavor.

It is not the place here to analyze or even evaluate Hasse's political opinions, expectations and deeds during the Nazi regime in Germany. This would make sense only after carefully scrutinizing, without prejudice, all the relevant documents of the time. Hasse himself has contributed to this by depositing his complete *Nachlass* at the University Library in Göttingen where it is accessible to historians. Certainly Hasse believed what he wrote to Davenport, *"that reason will come back in due course"*.[6]

9.3 Deuring's Letter to Hasse

We have seen in Sect. 9.2.1 that against Hasse's recommendation, Deuring had been rejected for a position as *Dozent* in Göttingen, because of political opposition against Hasse from the side of the dominant Nazi functionaries in the faculty committee. In view of Deuring's mathematical capacity this was a blatant mistake of the committee. Already on 9 May 1936, 5 months after this negative decision, Deuring wrote to Hasse announcing a manuscript which contained new and important ideas how to attack the RHp in the case of higher genus $g > 1$.

In fact, Deuring had found the algebraic setting which seemed suitable for carrying the proof. He provided an algebraic construction of the endomorphism ring

---

[6]That was what apparently many people at those times still expected or at least hoped. And not only at those times.

of the Jacobian of an algebraic curve, without explicitly regarding the Jacobian as an algebraic variety of dimension g. He worked directly within the function field of the curve, where the Jacobian is situated as the divisor class group of degree 0. Deuring created the algebraic notion of what had been called "correspondence" in classical complex analysis and algebraic geometry. (Here, the word "correspondence" is to be understood as a mathematical term, like "field" or "group" etc. It should not be confused with the meaning of "correspondence" in colloquial language, where it means, e.g., the exchange of information by means of letters. I hope that there will not arise misunderstandings.)

Let me cite from Deuring's letter:

"Dear Mr. Hasse, in the recent weeks I have tried to generalize your results for elliptic function fields to fields of higher genus. In this endeavor I have been successful up to the construction of the multiplier ring and the proof that it is algebraic. Perhaps you have already more results on this topic, therefore I am sending you the introduction of a paper which I am planning. There, the algebraic results are stated only. The complete proofs are already obtained but are still in a monstrous state."

Deuring uses the name "multiplier" for what nowadays is called "endomorphism" of the Jacobian. In this he followed Hasse and his team who had used "multiplier" in the elliptic case.

The proof that the "multiplier ring" is algebraic (i.e., that every element satisfies an algebraic equation over $\mathbb{Q}$) does not appear in Deuring's paper [Deu37]. Perhaps there was a flaw in Deuring's original proof which he was not able to correct at the time ?

Hasse answered 2 days later. He was on the brink to travel to Königsberg to give a colloquium talk there and hence had not yet been able to look at the details of Deuring's manuscript, but he already had a glimpse into it. Hasse wrote:

"... In any case I am sure that you have now obtained the base for handling the Riemann hypothesis in arbitrary function fields. I am convinced that I will be able to obtain a proof of the RHp, linking your results with my own Ansatz which I have been thinking about during the past weeks. I plan to look after this as soon as possible, also with respect to streamlining your proofs."

It appears that Hasse, although he wrote he *"had been thinking about it"*, did not yet have *"more results"* as Deuring suspected. In any case Hasse fully acknowledged Deuring's achievement.

There followed an exchange of several letters between Deuring and Hasse. Deuring sent the first part of his paper for publication and Hasse, as he usually did with submitted papers, went over it, editing and simplifying the manuscript. For this he enlisted the help of his assistant H.L. Schmid. The paper was published 1937 in Crelle's Journal [Deu37]. It appears that the editing had been quite extensive, in particular the (well written) introduction shows unmistakably Hasse's style. Unfortunately I did not find Deuring's original manuscript and so I have not been able to compare both versions. A second paper followed later in 1940 [Deu40].

9.3.1   Correspondences

Now, what were the new ideas of Deuring?

In a sense, Deuring's ideas were not "new" but rather they are to be regarded as a continuation of Hasse's ideas in the same vein. He generalized Hasse's "Step 1" (see page 91), i.e., the algebraic construction of the endomorphism ring of the Jacobian of a function field, now for the case of arbitrary genus $g \geq 1$.

In the elliptic case, Hasse had worked with the "double field" $\mathcal{F} = FE$, the compositum of two algebraically independent but isomorphic function fields $F|K$ and $E|K$, the base field K being algebraically closed.[7] Deuring too worked with $\mathcal{F}$ but he did *not* assume that $F|K$ and $E|K$ are isomorphic. For, it turned out that the theory of correspondences $J(E) \to J(F)$ can be developed for arbitrary function fields $F|K$ and $E|K$ without assuming that they are isomorphic. The only difference is that in this general case the correspondences do not form a ring but a group only; let us denote it by $\mathsf{M} = \mathsf{M}(E, F)$. The group operation in M may be written as addition or multiplication.

More precisely let

$$\mathcal{F} := FE = \mathrm{Quot}(F \otimes_K E) \,.$$

$\mathcal{F}$ is considered as a function field with base field E. Like Hasse, Deuring works with the transcendental primes M of $\mathcal{F}|E$. The residue map modulo M induces an isomorphism $\mu : F \to \mathcal{F}M$, and we have the situation of (7.15) on page 97.

But now, unlike Hasse, Deuring considers transcendental primes M of arbitrary degree $[\mathcal{F}M : E]$ whereas Hasse in the case $g = 1$ got away with transcendental primes M of degree 1, i.e., for which $F\mu \subset E$. In Deuring's more general situation consider the following maps of divisors which are indicated by arrows:

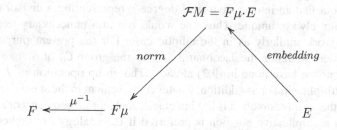

If M is of degree 1 this diagram reduces to the diagram on page 94.

There results a homomorphism of the divisor group $\mathrm{Div}(E|K)$ into $\mathrm{Div}(F|K)$, composed of three steps as indicated in the diagram. Accordingly every divisor B of $E|K$ is mapped as

$$B \mapsto N_{\mathcal{F}M|F\mu}(B) \cdot \mu^{-1} \,. \tag{9.1}$$

---

[7] As to the terminology and the description of the situation see page 96 ff.

This looks similar to the formula in the elliptic case (7.11) on page 94. But there $F\mu \subset F$ whereas here B has first to be considered as a divisor of the extension $\mathcal{F}M$ before the norm operator is applicable. Note that the divisor group of $E|K$ embeds injectively into the divisor group of the finite extension $\mathcal{F}M|K$ and will therefore be considered as a subgroup of the latter. (Some people use the name "conorm" for this injection.)

As in the elliptic case, the map (9.1) is denoted by μ again but as left operator. Thus μB is the image of the divisor B of $E|K$ in the map (9.1). This map is called a "correspondence" from E to F. The map preserves divisor equivalence and divisibility of divisors. However it does *not* preserve the degree as it does in the elliptic case (7.11) where one works with primes M of degree 1 only. In the present situation the degree of B is multiplied by the field degree $[\mathcal{F}M : E] = \deg M$, for this happens in the process of embedding whereas the norm operator preserves the degree. That is, we have

$$\deg \mu B = \deg M \cdot \deg B .$$

Here the degrees are to be understood in the function field of the respective divisors: μB is a divisor of $F|K$ while M is a divisor of $\mathcal{F}|E$ and B is a divisor of $E|K$.

If $\deg B = 0$ then $\deg \mu B = 0$, hence μ induces a homomorphism of the divisor class groups of degree 0, i.e., the Jacobians:

$$\mu : J(E|K) \longrightarrow J(F|K) . \tag{9.2}$$

Note: In the elliptic case, Hasse had introduced the Jacobian of $F|K$ as an additive group, consisting of the prime divisors P of $F|K$ with the addition given by (7.10) on page 92. In the case of higher genus $g > 1$ one could try to do the same by using the integer divisors of degree g instead of the prime divisors. But it turns out that an integer divisor of degree g representing a divisor class of degree 0 is not always unique. Thus one would run into unnecessary troubles if one tries to work similarly as in the elliptic case. For the present purpose it is more convenient to *define* the Jacobian J just as the group $\mathcal{C}_0$ of divisor classes of degree 0, as we have done in (9.2) already. The group operation in J may be written as multiplication or as addition, whatever convenient in the special situation. Historically the additive notation is used in classical algebraic geometry or analysis. Whereas the multiplicative notation is preferred if the analogy to number theory should be pointed out. For the time being I shall stick to the multiplicative notation which was used by Hasse and Deuring.

I am also writing $M(B)$ for μB. For an algebraic prime divisor P of $\mathcal{F}|E$ Deuring defines

$$P(B) = P^{\deg B} \tag{9.3}$$

(in multiplicative notation). By linearity *every* divisor $A \in \mathrm{Div}(\mathcal{F}|E)$ is now represented as a homomorphism $B \mapsto A(B)$ from $\mathrm{Div}(E|K)$ to $\mathrm{Div}(F|K)$.

Restricted to the Jacobians (= divisor class groups of degree 0) we find $\text{Div}(\mathcal{F}|E)$ represented as a group of homomorphisms $J(E) \to J(F)$. These homomorphisms are also called *correspondences* and the maps belonging to the transcendental primes M are now called *prime correspondences*.

From (9.3) we see that an algebraic prime P acts trivially on the Jacobian $J(E)$ since there $\deg(B) = 0$. It follows that every algebraic divisor, i.e., a divisor which is composed of algebraic prime divisors only, also acts trivially on $J(E)$. Deuring shows that principal divisors too act trivially on $J(E)$, i.e., we have

$$A \sim 1 \Longrightarrow A(B) \sim 1 \qquad \text{for all divisors } B \text{ of } E|K. \tag{9.4}$$

Perhaps it is not without interest to point out the reason for this. It is sufficient to verify this when $B = Q$ is a prime divisor of $E|K$. The prime Q extends uniquely to a prime divisor of $\mathcal{F}$ which is trivial on F. The extended prime will also be denoted by Q. The residue map modulo Q maps $\mathcal{F}|E$ onto $F|K$ and as such is a *good reduction* of $\mathcal{F}|E$. Together with this goes a divisor reduction map Q : $\text{Div}(\mathcal{F}|E) \to \text{Div}(F|K)$ which preserves degree and equivalence of divisors. Let us denote for a moment the image of a divisor $A \in \text{Div}(\mathcal{F}|E)$ by $\overline{A}$. Then

$$A \sim 1 \Longrightarrow \overline{A} \sim 1.$$

Now, if M is a transcendental prime divisor of $\mathcal{F}|E$ then the definition of its reduction $\overline{M}$ coincides precisely with $M(Q) = \mu Q$ as defined by (9.1). And similarly for an algebraic prime divisor in view of (9.3). It follows that for any divisor $A \in \text{Div}(\mathcal{F}|E)$ we have $A(Q) = \overline{A}$. Hence (9.4).

I have said already that the theory of good reduction had been systematically developed later only. But in Deuring's paper [Deu37] he already performed the necessary computations which later he systematized in his reduction theory [Deu42]. By the way, Deuring used the terminology "regular reduction" for what today is called "good reduction".

Two divisors A_1, A_2 of $\mathcal{F}|E$ are called *coarsely equivalent* if they are equivalent in the ordinary sense up to an algebraic divisor. We have seen that the action of A as a correspondence $J(E) \to J(F)$ depends only on the coarse equivalence class of A.

The main result of Deuring is:

The group of coarse equivalence classes acts faithfully as correspondences $J(E) \to J(F)$, i.e., it is isomorphic to $\mathsf{M}(E, F)$.

$\mathsf{M}(E, F)$ does not contain all group homomorphisms of $J(E)$ into $J(F)$ in the abstract sense. It contains precisely those homomorphisms $J(E) \to J(F)$ which are generated by prime correspondences as defined above. In the classical case when the base field $K = \mathbb{C}$ is the complex number field, then M coincides with the group of correspondences in the sense of algebraic geometry. But this interpretation is irrelevant in the algebraic setting of Hasse and Deuring. The main point in Deuring's paper is to establish an algebraic theory of correspondences, valid in *every* characteristic. He refers to an old paper by Hurwitz [Hur86] where correspondences of Riemann surfaces are investigated from the analytic point of view, and he says that this had inspired him to translate it into the algebraic setting. In the introduction of his paper [Deu37] he also refers to Severi's book [Sev26] where correspondences are treated within classical algebraic geometry. But Deuring's theory was conceived independently, and the reference to Severi was inserted afterwards only, after Weil and Lefschetz had pointed this out to Hasse. (See page 150 below.)

Recall that in the 1930s algebraic geometry in characteristic $p > 0$ had not yet been established, at least not in a form which could have been used for Hasse's problem of RHp.

REMARK If E is isomorphic to F then, by means of a fixed such isomorphism M can be regarded as an (additive) group of operators on $J(F)$. As such it is in fact a ring, the product being the successive application of two correspondences. One has to prove that the successive application of two correspondences is again a correspondence in the sense defined above. Deuring does this but the proof is somewhat involved, different from Hasse's case of genus 1 where correspondences are given by meromorphisms of the function field.

9.4 Hasse's Letter to Weil

In the year 1936 there were a number of letters exchanged between Weil and Hasse, about varying topics.[8] Here we are interested in those letters in which the RHp was discussed. One of these was Hasse's letter of 12 July 1936, written as a reply to Weil's of 8 July where Weil had written:

> "*Finally I did neither go to the Caucasus nor to Oslo, mainly because I want to work quietly on my mémorial-book about group theory which should have been completed long ago... Please do inform me occasionally about the number theoretic news which you will hear in Oslo!*"

Weil refers to earlier letters where he had told Hasse that he plans to do a hiking tour in the Caucasus together with Delaunay. On the way he would pass Göttingen and would like to visit Hasse. Later he wrote that he could not realize the appointment with Delaunay but he would meet Hasse at the International Congress of Mathematicians in Oslo which was scheduled for July 1936. Now, in the present letter he regrets that the meeting with Hasse will not be possible, neither in Göttingen nor in Oslo, but he is asking for information about number theoretic news from there.

---

[8]Among them was the work of Weil's student Elizabeth Lutz at Strassbourg on the structure of the group of rational points of an elliptic curve over a p-adic field (p-adic uniformization). Weil had proposed to have Lutz's paper published in Göttingen as a "*sign of continued cooperation*". Hasse gladly agreed; he accepted Lutz's paper for Crelle's Journal [Lut37].

Hasse replied immediately, one day before his departure to the Oslo congress which started on 13 July. Instead of news from Oslo he sent number theoretic news from Göttingen.

Indeed, in Hasse's seminar and in the *Arbeitsgemeinschaft* (workshop) of the summer semester 1936 in Göttingen, quite a number of important new results had been obtained. Several of these are documented in volume 176 of Crelle's Journal which contains twelve papers of high level of the number theory group around Hasse, among them Ernst Witt, H.L. Schmid, Martin Eichler and Oswald Teichmüller. It appears that Hasse was quite proud about this and he wanted to communicate it to his correspondence partners. While Hasse's efforts to *"rebuild Göttingen as a high-ranking mathematical site"* had failed on a larger scale, he had been successful with his small algebra group of young people. In his letter to Weil he reports only on three of those new results:

"First of all you will certainly be interested to know that Mr. Witt has now proved the functional equation for the L-series in congruence function fields. He used a very nice analogue of the classical proof with theta functions. I am enclosing a short sketch of his proof."

For the terminology of "congruence function fields" see page 16. When Hasse mentions L-series he refers to L-series $L(\chi, s)$ in the sense of class field theory, belonging to divisor class characters χ. These are not to be confused with F.K. Schmidt's L-polynomial for which the functional equation had been already established, see Sect. 4.4.

Hasse continues:

"Furthermore, I myself have explicitly determined the group of p^n-primary numbers of a p-adic number field. I introduced a generalized notion of exponentiation and applied the beautiful theory of Witt on the cyclic generation of degree p^n in characteristic p. You will find a sketch of this on a second sheet."

Here Hasse refers to Witt vectors which are also among the results achieved in the Göttingen *Arbeitsgemeinschaft*. Witt's paper appeared in [Wit36]. It seems not to be widely known that Witt vectors were first encountered by Hasse's Ph.D. student H.L. Schmid on the occasion of discovering explicit reciprocity laws for cyclic extensions of degree p^n of function fields in characteristic p. See [Sch35b]. Witt's contribution was the introduction of the so-called ghost-components by which the computations become much clearer and better to handle, thus universally applicable also in other mathematical situations.

Hasse's result on p^n-primary numbers was indeed a very important result in connection with explicit formulas for the reciprocity law. In the 1920s Artin and Hasse had eagerly searched for it but without success for $n > 1$. See [FR08]. But now, with the theory of Witt vectors at hand, Hasse had found a solution [Has37b]. In later years Shafarevich took up the case [Sha51] and, based on Hasse's result, obtained a general explicit reciprocity law as a comprehensive addition to Part II of Hasse's class field report [Has30].

Finally Hasse mentions what interests us in the present context:

"Finally, Deuring has had the crucial idea which led to the generalization of my theory in the elliptic case to arbitrary genus $g \geq 1$. It is the algebraization of the theory of correspondences between two algebraic function fields, hence not, as I always had thought, using the field of abelian functions which has g variables."

And Hasse proceeds with a sketch of Deuring's main ideas and methods, similarly as I have reported it in the foregoing section. He closes this with the comment:

"In order to reach the proof of the Riemann hypothesis from here, two items are still missing: First, a theory describing the behavior of the differentials of the first kind under these correspondences. Secondly, the generalization of the theory of the norm which here is related to the field degree $[F\mu \cdot E : F\mu]$."
(The notation is ours, it refers to the diagram on page 143.)

When Hasse here refers to his theory of the "norm" he means his norm function which he had defined in the elliptic case as $\mathcal{N}(\cdots) = [F : F\mu]$, see page 98. Hasse ends this part of his letter as follows:

"Since I got to know Deuring's theory only very recently I have not yet found time to think over these two generalizations. In any case it is roughly seen, also in this general case, that the Riemann hypothesis is reduced to the abstract analogue of the fact that the well known Hermitian form belonging to the period matrix is positive definite."

A similar text is found at the end of Hasse's report on his lecture at the ICM in Oslo [Has37d].

Thus Hasse was giving the direction for the work ahead:

Problem *Find a positive definite quadratic form on the endomorphism ring* M *of the Jacobian of a function field* $F|K$, *which is the algebraic equivalent of the positive definite Hermitian form in the classical situation.*

9.5   Weil's Reply and Lefschetz's Note

Hasse's letter to Weil was quite long: 8 typewritten pages plus some handwritten notes. As said above, Hasse wrote it just before his departure to the International Mathematical Congress in Oslo. Upon his return he found Weil's answer in his mail, dated 15 July 1936, thus 5 days after Hasse had dispatched his letter. Weil's answer was also long, about 5 pages. He concentrated on Deuring's work on correspondences, he did not comment on the other information which Hasse had sent him. He pointed out the connection of Deuring's theory with what was known in Italian algebraic geometry. Weil's letter started as follows:

"Dear Mr. Hasse, I have read your letter and the enclosed notes with greatest interest. As you may already have guessed, I was particularly pleased to see the generalization of your theory of elliptic function fields; and it is fine that due to Deuring's idea the solution of that problem seems to be imminent . . . It is a very fortunate idea to consider the singular correspondences in order to generalize the algebraic theory of complex multiplication . . . "

In the former correspondence between Hasse and Weil, the Jacobian J appeared as a g-dimensional abelian variety. This implied that for the RHp one first would have to develop the theory of abelian varieties of higher dimension in characteristic p, including their divisors as the analogue of the ϑ-functions, before turning to the construction of the ring of their endomorphisms. But here we see that Weil acknowledges the change Deuring's idea had brought about in order to avoid the theory of abelian varieties in characteristic p. Now the algebraic construction of the endomorphism ring of the Jacobian was possible by means of Deuring's correspondences.

But, Weil continues, several of the main ideas for this theory had already been developed. The transcendental theory of correspondences had been provided by Hurwitz in his well known paper of 1886 [Hur86]. (This is the same paper which Deuring had cited as his source for inspiration for his algebraic theory.) Thereafter the theory had been taken up by the Italian geometers, in the framework of the old geometry but in a truly algebraic spirit. Weil refers to Severi's *"Trattato di geometria algebrica"* [Sev26]. The consideration of the field $\mathcal{F} = FE$ is classical in this connection, he writes. Weil points out that the whole of Deuring's theory can be found in Severi's book, in Chapter VI, §§ 60–71, but in the framework of classical algebraic geometry. He continues:

"Certainly I know how necessary and sometimes how difficult it is to translate existing results of this kind into the language of modern algebra. But, in my opinion, it is important in such investigations never to neglect the connection with the older theories . . . mainly in order not to throw away irreplaceable guidelines. I believe that this will also prove true with respect to the problem at hand."

Weil adds some general remarks. First, he does not see any discrepancy between the two viewpoints: that of the correspondences on the one hand, and of the field of abelian functions on the other hand. He refers to Hurwitz [Hur86] who already knew that relations between the periods of abelian functions are related to the existence of singular correspondences. This, he writes, will easily lead to the solution of the first open problem which Hasse had stated in his letter, namely the representation of the correspondences on the space of differentials of the first kind. As to the second problem in Hasse's letter, namely the behavior of the norm function, Weil believes that the norm can *only* be defined via the field of abelian functions.

REMARK This is not in contradiction to Hasse's statement that the norm is related to the field degree $[F\mu \cdot E : F\mu]$ (see page 148). Whereas Hasse's "norm" refers to the envisaged positive definite quadratic form on M (sometimes one speaks of a "normed space"), Weil's "norm" refers to what today is called the degree of an isogeny of the Jacobian variety of the curve. As it turned out, Hasse's statement is the one which is essential for the proof of RHp, see Sect. 10.2, in particular formula (10.27) where $d'(M)$ is the "norm" in the sense of Hasse. In the elliptic case both these "norms" coincide.

But Weil was not the only one who saw the connection to algebraic geometry. At about the same time Hasse received a note from Lefschetz with similar content. Hasse had met Lefschetz in Oslo and they had a discussion about Deuring's theory. The following note was apparently written by Lefschetz right after the Congress while both were still in Oslo.

> "Oslo, den 20–VII–1936 Dear Prof. Hasse:
> I have thought about the Theorem of Deuring which you communicated to me, and am now almost certain that it is quite old. It is probably implicit in some of Severi's early papers, and explicit in some paper of Carlo Rosati who devoted his whole work to algebraic correspondences. I am certain that I could give you the exact reference upon my return to Princeton. I should say that it is at least 15 years old. It may even be implicit in the work of A. Hurwitz. Deuring's result should not be published before he has the opportunity to verify this. It is n years since I have dealt with these questions, hence I cannot tell you much more now. I am certain however that I could prove the theorem very rapidly even now.
> Auf Wiedersehen Lefschetz
> We are leaving Tuesday morning $\longrightarrow$ Bergen."

With this information by Weil and by Lefschetz at hand Hasse immediately informed Deuring who replied on 10 August 1936:

> "I have at once checked with Severi's book. Weil (and hence Lefschetz) are completely right. Everything can be found quite explicitly in Severi, however with completely different proofs. But it appears to me that my proofs are not superfluous. For they show that it is possible to approach those problems and solutions also from the arithmetic viewpoint (not to mention the fact that the old proofs have to be checked whether they are valid also in the general case [in arbitrary characteristic] which however I do expect). But first I have to familiarize myself with it."

As a consequence, Deuring in the introduction of his paper [Deu37] added a reference to Severi's book, but he pointed out that his paper was written independently. (I have mentioned this already on page 146.)

Later, in September of that year Hasse received a letter from his former student O.F.G. Schilling who at that time studied in Princeton at the Institute for Advanced Study. Schilling participated in Lefschetz's seminar on algebraic geometry at

Princeton University, and Lefschetz had asked him to transmit to Hasse some references to papers on the theory of correspondences in Italian algebraic geometry. Schilling wrote on 25 November 1936:

"During a preliminary discussion for a seminar on algebraic geometry Professor Lefschetz told me that in Oslo he had talked with you about certain new results by Mr. Deuring on the theory of complex multiplication. Mr. L. was of the opinion that the definition of the commutator algebra of a Riemann matrix belonging to a given Riemann surface with the help of algebraic correspondences is already known ... See the well known paper by Hurwitz. By chance I found today an article by Mr. L. in which those questions are discussed, with references to literature ...I would like to mention that I am sending this bibliographic reference at the instigation of Professor Lefschetz ..."

Schilling means Lefschetz's article [Lef28] on correspondences of curves.

Hasse in his reply to Schilling explained to him the importance of Deuring's construction. For, Deuring's results are obtained algebraically without using the integrals of the first kind, and hence valid in arbitrary characteristic. He closes his letter with the words:

"I hope that Mr. Lefschetz will revise his decree "very old" when he reads Deuring's paper in detail."

Well, I have some doubts whether Hasse's hope did realize. I have no information that Lefschetz even had looked at Deuring's paper. In the "MacTutor History of Mathematics archive" I have found the following text about Lefschetz:

"... if something was to him clearly true, he would consider it at best a waste of time producing a rigorous argument to verify it."

SIDE REMARK Otto F.G. Schilling (1911–1973) had been a student of Emmy Noether in Göttingen. When Noether was forced to leave Göttingen she asked Hasse to take over. Hence Schilling went to Marburg where he got his doctoral degree in the year 1934 under the supervision of Hasse, with a thesis about the role of algebras in the theory of algebraic number fields [Sch35a]. In the summer of 1935 he stayed in Cambridge with Davenport and Philip Hall. Thereafter he went to the Institute of Advanced Study in Princeton where he was accepted as a member on the recommendation of Hasse to Hermann Weyl. From there he held contact with Hasse by mail until 1937. Schilling belonged to the well known Schilling family who since more than 100 years ran a bell foundry in the state of Thuringia (Germany). The Schilling bells were known worldwide due to the quality of their tone and harmony. In the year 1936 Otto had his 25 birthday and at this point in his life he had to decide whether to follow the family tradition and join the business of his father, i.e., bellfounding. There is a letter of Schilling from Princeton to Hasse, dated 13 February 1936, asking for his advice on the matter. He would like to stay with mathematics, but did Hasse believe that his, Schilling's, mathematical skills would be sufficient for an academic career? Of course, Hasse in his reply could not give a definite advice but wrote a somewhat conditional recommendation to stay with mathematics. This helped Otto Schilling to decide his way. (The family business of bell founding was taken over by Otto's brother.) In Princeton he met A.A. Albert who in 1939 got him a position as professor at the University of Chicago. Schilling's book on valuation theory [Sch50] was the first book which contained a systematic theory of general valuations in the sense of Krull. 1961 Schilling accepted an offer from Purdue University where he remained until his retirement.

9.6   The Workshop on Algebraic Geometry

Both letters, by Weil and by Lefschetz, pointed to classical algebraic geometry where Hurwitz's theory of correspondences had been further developed. These two hints made Hasse wish to look more closely into the concepts and methods of algebraic geometry, in order to find ideas how to construct the positive definite quadratic form which, as he had written to Weil, he was looking for (see page 148).

Classical algebraic geometry (i.e., over the complex numbers as base field) had seen a fast development in the first decades of the twentieth century, notably by Italian geometers such as Castelnuovo, Enriques, Rosati, Severi. But by the 1930s this "Italian" algebraic geometry had somewhat come into disrepute. The proofs were criticized to have no precise foundation, so that the arguments used could not always be checked. The point was not that there were some erroneous results produced; errors do occur not infrequently in mathematical treatises. But it is strictly required that in a mathematical publication the arguments used in proofs should be clear and susceptible to checking by any person with sufficient mathematical knowledge.

Well, by writing down these words I get some doubts whether this is indeed so in reality. When we read a mathematical paper then it is necessary also to understand what the author does *not* say but tacitly assumes the reader to be acquainted with, namely certain prerequisites which have not been spelled out. Whenever the author says "obviously" or "it is easy to see" or something of this sort, we are invited to accept some facts, unspecified by the author, and to apply them in the situation at hand. And many times we have to do this even without the author mentioning anything at all in this respect. In view of this, the relation between the Italian algebraic geometry and the general mathematical scene in the 1930s may to some extent be described as a "clash of cultures of scientific publishing".[9] The desire to overcome this situation and to develop a universally accepted language in particular for algebraic geometry was beginning to rise. But at that time there were only few people who were able and willing to start this job. (For instance van der Waerden, Zariski and later A. Weil.)

---

[9] I have found this expression in Schappacher's survey [Sch07].

In any case, Hasse too seems to have had difficulties with the way in which Italian algebraic geometry was presented. He wished to understand what Severi in his book [Sev26] had said about correspondences of curves. As I have already indicated, never before Hasse had met any occasion to work in algebraic geometry. So he planned a small conference on algebraic geometry, today we would say "workshop". As he wrote to Blaschke on 24 December 1936:

> *"Originally this* [the conference] *was planned as a comfortable method for us ignorant people in Göttingen to be introduced into the various branches of algebraic geometry. Therefore we have invited the main representatives of the various branches and asked them for <u>introductory lectures.</u>"*

But he adds what could be said about so many "workshops" even today:

> *"However, everybody wished rather to talk about recent research results and hence our program has become somewhat different from what we originally intended."*

The first choice of speaker on algebraic geometry was of course van der Waerden who at that time held a position as full professor in Leipzig. During the last years he had written, originally inspired by Emmy Noether, quite a number of papers with important contributions to the foundation of algebraic geometry. (See, e.g., [vdW83].) But van der Waerden appears to have been quite busy at that time. He would only come to Göttingen if this travel could be combined with a visit to Hamburg where he had been invited too. Finally, on 2 December 1936 he wrote to Hasse:

> *"My talk in Hamburg is scheduled for Saturday 9 Jan. Hence I would be able to talk in Göttingen one of the days before that, i.e., Wednesday, Thursday or Friday…I would like to talk about the general notion of multiplicity, in particular of points of intersection, and also about multiple points on algebraic surfaces."*

In accordance to van der Waerden's wish Hasse then timed the Göttingen workshop for Wednesday 6 January to Friday 8 January 1937.

I have not found any document showing the program of this conference. But from various letters of Hasse and his colleagues I was able to extract the following information about the speakers and the topics of their talks: (These may not have been the final titles.)

1. Van der Waerden: General notion of multiplicity, in particular of points of intersection. (2 h)
     Modules of algebraic curves. (2 h)
2. H.E.W.Jung: Arithmetic theory of algebraic functions of two variables. (4 h)
3. H. Geppert: Introduction to the way of thinking in Italian algebraic geometry. (4 h)
4. M. Deuring: Algebraic theory of correspondences. (3 h)

Hasse had also invited W. Krull who at that time held a position at the University in Erlangen. Krull had worked on what today is called "commutative algebra"; his survey on ideal theory had become a very influential book [Kru35]. But he could not come to the Göttingen workshop although he was very interested in the program. Also, André Weil wrote on 21 December 1936 that he would gladly try to come to the conference in Göttingen but just on 6 January 1937 he had to board the boat for the USA where he intends to stay at the Institute for Advanced Study for one year.

In addition to the invited speakers there participated a number of young and brilliant people at the conference. Blaschke announced in a letter to Hasse that he will be accompanied by Zassenhaus, Bol, Maak and Petersson. From Göttingen there were in all probability the participants of Hasse's seminar and the *Arbeitsgemeinschaft* present, in particular Witt, H.L. Schmid and perhaps Behrbohm. I do not know whether Teichmüller participated. In January 1937 he was still in Göttingen, changing to Berlin later in April 1937. But his interests at that time were directed towards the theory of conformal functions where he was to give essential contributions. We should also note that Hasse recently had been able to get H. Rohrbach, as assistant professor to Göttingen[10] hence probably he participated too. Similarly Hanna von Caemmerer (later Hanna Neumann) who at that time studied with Hasse as her Ph.D. advisor. [11]

One can see from these names that there were good prospects for the conference to be successful, in the sense that there may have emerged useful ideas pointing the way to the construction of the envisaged quadratic form, hence to the proof of the RHp. After all, van der Waerden was a specialist for the notion of intersection multiplicity, which turned out to be an essential ingredient of the positive definite quadratic form which Hasse was looking for. Jung had worked all his life on function fields of two variables; note that the field $\mathcal{F} = F \cdot E$ of Hasse and Deuring is of that kind. Geppert knew Italian well (for family reasons) and had close contacts to the general Italian mathematical scene; he worked on algebraic geometry and was well acquainted with the way of doing algebraic geometry in Italy. And there were several young and interested participants who may have taken up the torch and proceed to the goal.

However, in the literature of 1937 and the following years I could not find anything new of substance referring to this conference. According to my present knowledge the Göttingen conference had not been helpful for the RHp.

---

[10]Rohrbach had held a position at Berlin University but had got there into trouble because of political reasons.

[11]She too came from Berlin. There she did not like the Nazi dominated atmosphere and therefore she had asked Rohrbach where to go for her Ph.D. Rohrbach suggested to come to Göttingen too and study with Hasse. The latter proposed to her to work on the foundation of arithmetics in higher dimensional function fields. But one year later von Caemmerer suddenly left Göttingen and went to Britain where she married Bernhard Neumann with whom she had been secretly engaged in Berlin already. He was Jewish and therefore had to leave Germany. Both Hanna and Bernhard grew to become well known group theorists.

Perhaps Hasse had expected from this conference a similar outcome as he had experienced 6 years before, in February 1931, with his workshop on algebras. At that workshop he had brought together mathematicians who had been active in the theory of algebras, of class fields and group representations, in order to join forces to attack his main conjectures on cyclic simple algebras over number fields. This had been successful insofar as immediately after that workshop he had succeeded to prove the local-global principle for cyclic algebras and thus complete his theory of cyclic algebras over number fields [Has32]. And some time later he could show the cyclicity of arbitrary simple algebras over number fields, jointly with Emmy Noether and Richard Brauer who had participated in that workshop. I have told that story in [Roq05].

But nothing like this emerged from the Göttingen workshop on algebraic geometry in 1937. It appears that there was nobody who was able and willing to continue what Deuring had started, namely translating the mathematical language used in Italian algebraic geometry into the language of algebra used by Hasse and his work group.

In fact, the work needed was like a translation from one (mathematical) language into another, i.e., from the language of algebraic geometry into that of algebraic number theory. André Weil has expressed this in his beautiful letter which he wrote to his sister in 1940 from the military prison "*Bonne Nouvelle*". The letter is published in [Wei79a]. Weil had been imprisoned as a conscientious objector to military service, and while waiting for the court trial he used his time there to think about attacking the RHp which he knew Hasse had not yet captured. (Later I shall report more about Weil's mathematical activity at that time, see Chap. 12.) In the letter Weil talks about the role of analogy and intuition as impetus of mathematical discoveries. He compares his work with the decipherment of a text in three languages:

1. Number Theory,
2. Riemann's Theory of (complex valued) Functions,
3. Algebraic Theory of Functions over a finite base field.

He did not explicitly mention Algebraic Geometry but from the context it can be seen that somehow he considered Algebraic Geometry as a development of Riemannian Geometry in the algebraic spirit.

If I may use Weil's picture of a text in three languages, then what I have reported until now was the translation of a problem from language (1), namely counting solutions of diophantine congruences initiated by Davenport and Mordell, into a problem from language (3), namely the RHp. This translation was started by Artin and continued by Hasse. What was needed now, i.e., in 1937, was to obtain a translation from (2) to (3). Deuring had started this but it was necessary to continue.

If only someone at the workshop had been willing and able to look into the Italian literature on correspondences, in particular into Severi's book [Sev26] which had been mentioned by Weil in his letter to Hasse, and if he would have been able to translate it into the mathematical language developed and used by Hasse and

Deuring—then the positive definite quadratic form which Hasse was looking for could well have been constructed right then and there, during the conference or shortly after. All the prerequisites for this were available.

Summary

Already in 1934 Hasse contemplated the possibility of generalizing his work to function fields of higher genus $g > 1$. A. Weil pointed out to him that this may perhaps be achieved by studying the Jacobian of a curve and their ϑ-functions. But there are no signs that Hasse did anything definite in this direction. On the one side he was still heavily absorbed by working on his second proof in the elliptic case and by establishing the algebraic theory of function fields in general. On the other side his work was severely impeded by political difficulties which he had to face after changing from Marburg to Göttingen in 1934. He even thought of leaving Göttingen after Toeplitz, in the year 1935, had been able to induce the University of Bonn to offer him a position there. But the ministry in Berlin did not give the necessary permission.

When Hasse in 1935 tried to have Deuring get a position in Göttingen, this was rejected heavily and made impossible by the Nazi colleagues and the university rector. Nevertheless, some months later Deuring in a letter to Hasse developed an algebraic theory of correspondences for function fields of arbitrary genus $g \geq 1$. This was a decisive step towards the RHp for arbitrary function fields. When Hasse informed Weil about this then Weil pointed out in his reply that similar ideas are to be found in Italian algebraic geometry, in particular he referred to Severi. Consequently Hasse wished to obtain more information about the Italian algebraic geometry; to this end he organized a small workshop in Göttingen where competent speakers should report on algebraic geometry. But it turned out that this workshop did not reach its goal, namely producing ideas how to construct algebraically the positive definite quadratic form which in the classical case is given by the commutators of the period matrices.

Chapter 10
A Virtual Proof

In this chapter I would like to interrupt the historic line in order to put into evidence what I just said, namely that the proof of RHp could have been found already in 1937, in the framework of the theory of function fields. I will present here such a proof. In principle it can be regarded as a translation of Severi's proof from the language of algebraic geometry into the language of algebra. But I will not use any knowledge of the terminology and results of algebraic geometry. I shall use those notions and facts from the theory of function fields which were available to and preferred by Hasse at the time of the Göttingen workshop which I have discussed above.

Since such a proof did not materialize at that time I have called it a "virtual" proof.

The essential notion to be borrowed from classical algebraic geometry will be the *intersection multiplicity* of two prime divisors on a surface. I am going to point out that this notion does have its analogue also in classical algebraic number theory, namely in Hilbert's definition of *different* and *discriminant* divisors. Certainly Hasse knew Hilbert's *Zahlbericht* where these notions had been defined and used.

10.1 The Quadratic Form

10.1.1 The Double Field

Let $F|K$ be a function field and $g \geq 1$ its genus. The base field K is supposed to be algebraically closed. In the spirit of Hasse and Deuring I choose an auxiliary function field $E|K$ to be specified later. (This will be done in Sect. 10.3 where $E|K$ is chosen to be isomorphic to $F|K$. But for a moment it will be useful not yet to specify E, this will simplify certain details in the proof.) I consider the "double

© Springer Nature Switzerland AG 2018
P. Roquette, *The Riemann Hypothesis in Characteristic* p
in Historical Perspective, Lecture Notes in Mathematics 2222,
https://doi.org/10.1007/978-3-319-99067-5_10

field" $\mathcal{F} = FE$, the independent compositum of the two function fields $F|K$ and $E|K$. (See Sects. 7.4.2 and 9.3.1.) This field $\mathcal{F}$ is considered as a function field of one variable over E, and as such it is a base field extension of $F|K$. This is indicated in the diagram by double lines.

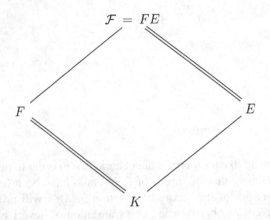

The divisor theory of this function field is based on those primes (=valuations) of $\mathcal{F}$ which are trivial on E. Following Deuring we have to distinguish two kinds of prime divisors of $\mathcal{F}|E$:

1. The transcendental primes M which are trivial on both E and F. The residue map modulo M induces in F an isomorphism into the residue field. Usually the residue map is normalized such that it is the identity on E. Then the residue field appears as a finite extension of E, generated by the image of F. The group generated by the transcendental primes is denoted by $\mathrm{Div}_{tr}(\mathcal{F})$. The divisors of $\mathrm{Div}_{tr}(\mathcal{F})$ are called "transcendental".
2. The algebraic primes. Every such prime belongs to a prime P of $F|K$, and it is the unique extension of P to $\mathcal{F}|E$. I denote that extension also by P, which means I identify the divisor group $\mathrm{Div}(F|K)$ with a subgroup of the divisor group $\mathrm{Div}(\mathcal{F}|E)$, as is usually done for base field extensions of function fields.

(Deuring [Deu37] speaks of "non-constant" and "constant" primes. Our terminology refers to the transcendence degree of the image field $FM|K$ or $FP|K$ which is 1 or 0 respectively.)

Thus the divisor group $\mathrm{Div}(\mathcal{F}|E)$ is the direct product of two groups:

$$\mathrm{Div}(\mathcal{F}|E) = \mathrm{Div}_{tr}(\mathcal{F}) \times \mathrm{Div}(F|K). \tag{10.1}$$

If the principal divisors are factored out then one obtains the group of (ordinary) divisor classes. If in addition the algebraic divisors $\mathrm{Div}(F|K)$ are also factored out then one obtains the *coarse* divisor classes in the sense of Deuring. The coarse equivalence of two divisors A and B is denoted by $A \approx B$. (In algebraic geometry the notation $A \equiv B$ is used.)

According to Deuring the group M of correspondence maps $J(E) \to J(F)$ of the Jacobians is isomorphic to the group of coarse divisor classes of $\mathcal{F}|E$. (See page 145). But here I need not use the theory of Deuring. Its role here is solely to provide us with a *motivation*. The following proof runs entirely within the group of coarse divisor classes, without any explicit reference to Deuring's theory of correspondences. I am going to construct a symmetric bilinear form $\sigma(A, B)$ on the divisor group $\mathrm{Div}(\mathcal{F}|E)$ with values in $\mathbb{Z}$—such that σ depends on the coarse classes of the divisors A, B only. It will turn out that the quadratic form belonging to σ is positive definite in the sense that

$$\sigma(A, A) > 0 \qquad \text{if } A \not\approx 1. \tag{10.2}$$

This is the quadratic form which Hasse had envisaged in his letter to Weil but did not succeed to find within his framework of divisor theory. (See page 148.) The following lines are to put into evidence that σ can well be constructed while consistently following Hasse's ideas. The positive definiteness of σ will lead to a proof of the RHp by applying it to the group of divisors generated by the Frobenius correspondence.

While working with coarse divisor classes we can forget the algebraic divisors in $\mathrm{Div}(F|K)$ since they are factored out anyhow. Accordingly, in view of (10.1) the group of coarse divisor classes is realized as the factor group of the group of transcendental divisors $\mathrm{Div}_{tr}(\mathcal{F})$, modulo its principal divisors. Here, a "principal divisor" of $\mathrm{Div}_{tr}(\mathcal{F})$ is obtained from an ordinary principal divisor $(x) \in \mathrm{Div}(\mathcal{F}|E)$ by cutting off its algebraic part. In other words: The principal divisor $(x)_{tr} \in \mathrm{Div}_{tr}(\mathcal{F})$ is given by

$$v_M((x)_{tr}) = v_M(x) \quad \text{if } M \text{ is transcendental} \tag{10.3}$$

$$v_P((x)_{tr}) = 0 \quad \text{if } P \text{ is algebraic}. $$

10.1.2 The Different

The main ingredient for the construction of the quadratic form σ is the notion of "different divisor". Consider two transcendental primes $M \neq M'$ of $\mathcal{F}$. Assume first that both are of degree 1 over E, then the residue maps mod M and mod M' induce isomorphisms $\mu \neq \mu'$ from $F|K$ into $E|K$. Hilbert had introduced a divisor $\mathfrak{D}(\mu, \mu')$ of $E|K$ which measures in some sense the "arithmetic distance" between two isomorphisms and which I call the "different" of μ and μ'. Its definition is given locally: For any prime Q of $E|K$ the multiplicity of Q in $\mathfrak{D}(\mu, \mu')$ is given by

$$v_Q \mathfrak{D}(\mu, \mu') := \min_x \{v_Q(x\mu - x\mu')\} \tag{10.4}$$

where x ranges over those elements in F for which $x\mu$ and $x\mu'$ are Q-integers.

It is straightforward to verify that this is $\neq 0$ for finitely many primes Q only, hence indeed (10.4) defines an integer divisor $\mathfrak{D}(\mu, \mu')$ of $E|K$. The condition that $x\mu$ is a Q-integer means $x\mu Q \neq \infty$ (in Witt's notation which I have explained on page 94, formula (7.12)). Thus x should be integer with respect to the prime μQ of $F|K$. Similarly for $\mu'Q$. The different $\mathfrak{D}(\mu, \mu')$ contains those primes Q of $E|K$ for which $\mu Q = \mu'Q$. For, if $\mu Q \neq \mu'Q$ there exists $x \in F$ with $x\mu Q = 0$ whereas $x\mu'Q = 1$, hence $v_Q \mathfrak{D}(\mu, \mu') = 0$ by definition. If $\mu Q = \mu'Q$ then $v_Q(\mathfrak{D}(\mu, \mu')) > 0$, and the minimum on the right of (10.4) is attained if x is taken as a prime element for μQ. (Recall that the base field K is supposed to be algebraically closed, hence there is no "inertia".) The relation $\mu Q = \mu'Q$ may be interpreted such that Q is "ramified" with respect to μ, μ'. In this sense the different $\mathfrak{D}(\mu, \mu')$ contains precisely the ramified primes.

Hilbert had given this definition in his "Zahlbericht". (See [Hil97], in particular §12 there.) Hilbert works in algebraic number fields, and μ, μ' are isomorphisms over a subfield. He did not talk about divisors but about ideals, as it was usual at that time. Actually, he called $\mathfrak{D}(\mu, \mu')$ an "element"; he then defined the usual different of a finite extension of number fields $F|F_0$ as the product of those "elements":

$$\mathfrak{D}(F|F_0) := \prod_{\mu \neq 1} \mathfrak{D}(1, \mu) \qquad (10.5)$$

where μ ranges over the isomorphisms of $F|F_0$ into its algebraic closure, and "1" stands for the identity isomorphism of F. Of course this method works also for finite separable extensions of function fields.

In the above situation (10.4) of two transcendental divisors I will also write $\mathfrak{D}(M, M')$ instead of $\mathfrak{D}(\mu, \mu')$. This symbol is now to be extended to arbitrary transcendental primes $M \neq M'$ of $\mathcal{F}|E$, which are not necessarily of degree 1. To do this, observe that the definition (10.4) can be given for arbitrary isomorphisms $\mu \neq \mu'$ of $F|K$ into the algebraic closure $\widetilde{E}$ of E. This yields a divisor $\mathfrak{D}(\mu, \mu')$ of a finite extension $E'|E$ which contains $F\mu$ and $F\mu'$. It doesn't matter which field since if $E' \subset E''$ then we identify the divisor group of $E'|K$ with a subgroup of the divisor group of $E''|K$ under the natural embedding. (Some people call this "conorm".) The isomorphisms of F into $\widetilde{E}$ correspond to the transcendental prime divisors of $\mathcal{F}\widetilde{E}|\widetilde{E}$, all of which are of degree 1 over $\widetilde{E}$. Thus $\mathfrak{D}(M, M')$ is now defined for any pair of different transcendental prime divisors M, M' of $\mathcal{F}\widetilde{E}|\widetilde{E}$. We extend this symbol bilinearly to $\mathfrak{D}(A, B)$ where the arguments A, B are arbitrary transcendental divisors of $\mathcal{F}\widetilde{E}|\widetilde{E}$ with the stipulation that they do not have a common prime. Then $\mathfrak{D}(A, B)$ is a divisor of a suitable extension field E' of finite degree over E.

If the divisors A, B are already defined over E, i.e., if they are contained in the subgroup $\mathrm{Div}(\mathcal{F}|E) \subset \mathrm{Div}(\mathcal{F}\widetilde{E}|\widetilde{E})$, then it is seen that their different $\mathfrak{D}(A, B)$ is a divisor of $E|K$. For instance consider a transcendental prime M of $\mathcal{F}|E$. In $\mathcal{F}\widetilde{E}|\widetilde{E}$ it splits into a product of primes of degree 1:

$$M = M_1 \cdots M_m \qquad (10.6)$$

where $m = \deg(M)$. The M_i are precisely the conjugates of M_1 over E, each appearing as often as the inseparability degree of the residue field of M over E indicates. If M' is another transcendental prime of $\mathcal{F}|E$ with the decomposition

$$M' = M'_1 \cdots M'_{m'} \tag{10.7}$$

then

$$\mathfrak{D}(M, M') := \prod_{i,j} \mathfrak{D}(M_i, M'_j) \tag{10.8}$$

is indeed a divisor of $E|K$ since it is invariant under conjugation over E and also the inseparability degree is taken care of.

Thus for all transcendental divisors $A, B \in \operatorname{Div}_{tr}(\mathcal{F}|E)$ the different $\mathfrak{D}(A, B)$ is now established as a divisor of $E|K$, provided that A, B do not have a common prime divisor. Taking the degree gives a symmetric bilinear symbol

$$\chi(A, B) := \deg_{E|K} \mathfrak{D}(A, B) \tag{10.9}$$

with values in $\mathbb{Z}$.

Our next step is to modify $\chi(A, B)$ such that it becomes a quadratic form on the group of coarse divisor classes. To this end I introduce the "degree form" $\psi(A, B)$ on $\operatorname{Div}_{tr}(\mathcal{F})$ as follows. Observe that a transcendental divisor of $\mathcal{F}|E$ can also be regarded as a divisor of $\mathcal{F}|F$, with F as the base field. Hence every transcendental divisor carries two degrees: one over E which I have denoted by $\deg(A)$ and will now denote by $\deg_{\mathcal{F}|E}(A)$ if it appears necessary. The other is the degree over F which is denoted by $\deg_{\mathcal{F}|F}(A)$. For brevity I will write

$$d(A) = \deg_{\mathcal{F}|E}(A), \qquad d'(A) = \deg_{\mathcal{F}|F}(A). \tag{10.10}$$

The degree form $\psi(A, B)$ is defined for transcendental divisors A, B of $\mathcal{F}$ as

$$\psi(A, B) = d(A)d'(B) + d'(A)d(B). \tag{10.11}$$

With this terminology the following important lemma holds.

Lemma *If at least one of the transcendental divisors A, B of $\mathcal{F}$ is principal in $\operatorname{Div}_{tr}(\mathcal{F})$ then*

$$\chi(A, B) = \psi(A, B). \tag{10.12}$$

Consequently, the form

$$\sigma(A, B) := \psi(A, B) - \chi(A, B) \tag{10.13}$$

*depends only on the coarse equivalence classes of A and of B, i.e., σ appears
as a bilinear symmetric form on the group of coarse divisor classes of F.*

Recall that $\chi(A, B)$ had been defined under the restriction that A, B should not
have a common prime divisor. The same restriction carries over to $\sigma(A, B)$ in view
of its definition (10.13). But the Lemma now allows to wave this restriction, for
in any coarse divisor class there exist transcendental divisors which are relatively
prime to any given divisor.

It will turn out that the corresponding quadratic form $\sigma(A, A)$ is positive definite;
this will lead to the proof of RHp. But first I have to present to you a proof of the
above lemma.

Let $E'|E$ be a finite extension of degree m. If we consider the divisor $\mathfrak{D}(A, B)$ of
E as divisor of E' then its degree $\chi(A, B)$ is multiplied by m. Similarly $\psi(A, B)$ is
multiplied by m; this follows from (10.10) since $\deg_{\mathcal{F}|E}$ stays fixed and $\deg_{\mathcal{F}|F}$ is
multiplied by m. Hence if the relation (10.12) is proved in $\mathcal{F}E'$ then, after dividing
by m the relation follows in $\mathcal{F}$. In other words: For the proof of (10.12) we may
replace E by E'. Let us choose E' such that both A and B split into prime divisors
of degree 1 over E'. For simplicity let us change notation (during the following
proof) and write again E instead of E'. Thus:

It is sufficient to prove (10.12) *under the additional assumption that both A
and B split into primes of degree 1 over E.*

REMARK In this way I am reducing the proof to the case when the relevant primes
are of degree 1. This method has been applied already in the definition of the
different, and will again be applied below in the proof of positive definiteness.

Let $A = (x)_{tr}$ be principal in the group of transcendental divisors. We
decompose B into prime divisors of degree 1 over E. By linearity it suffices to
prove the Lemma for each of these prime factors. Let M be one of them. As above
I denote by $\mu : F \to E$ the isomorphism induced by the residue map modulo M.

Since the different had been defined locally, we have to start with local
considerations. Let Q be a prime of $E|K$. It induces in the subfield $F\mu \subset E$ a
prime which is the image of some prime P of $F|K$. In Witt's notation as introduced
on page 94 we have

$$P = \mu Q = N_{E|F\mu}(Q) \cdot \mu^{-1}.$$

Consider the valuation rings $\mathcal{O}_P \subset F$ and $\mathcal{O}_Q \subset E$ and their tensor product

$$\mathcal{O}_P \otimes \mathcal{O}_Q \tag{10.14}$$

which is to be understood over the base field K. This is an integrally closed subring
of $\mathcal{F}$, consisting of those elements $y \in \mathcal{F}$ which are integer for each transcendental
prime divisor as well as for P and for Q.

Every divisor $A \in \mathrm{Div}_{tr}(\mathcal{F})$ defines a (fractional) ideal of $\mathcal{O}_P \otimes \mathcal{O}_Q$, consisting of those elements $y \in \mathcal{F}$ for which

$$v_{M'}(y) \geq v_{M'}(A) \qquad \text{for each transcendental prime } M' \text{ of } \mathcal{F} \qquad (10.15)$$

$$v_P(y) \geq 0$$

$$v_Q(y) \geq 0.$$

Recall that the prime P of $F|K$ is identified with its unique extension as a prime of $\mathcal{F}|E$, and accordingly $v_P(\cdots)$ denotes in (10.15) the corresponding valuation of $\mathcal{F}$. Similarly, Q is identified with its unique extension as a prime of $\mathcal{F}|F$ and $v_Q(\cdots)$ denotes the corresponding valuation of $\mathcal{F}$.

This ideal is called the "ideal of multiples" of A and is denoted by $[A]_{P,Q}$. For these ideals we have the product formula, valid for arbitrary transcendental divisors A, B:

$$[AB]_{P,Q} = [A]_{P,Q} \cdot [B]_{P,Q} \qquad (10.16)$$

which can be verified without difficulty.

Well, it appears that the verification of (10.16) is not quite straightforward, hence I will sketch a proof. This proof is due to Richard Brauer who communicated the essential idea to me in a letter, back in the year 1953. (My own proof had been somewhat complicated.) I am presenting Brauer's proof here since I know that this was preferred by Hasse. In his Lecture Notes [Has68] Hasse called it "*R. Brauer's happy idea*".

If $\mathcal{O}_P \otimes \mathcal{O}_Q$ is replaced by $\mathcal{O}_P \otimes E$, neglecting the prime Q, then one obtains an integrally closed subring of the function field $\mathcal{F}|E$. Such ring is known to be a Dedekind ring. The ideal of multiples of A in this ring is described by the first two conditions in (10.15), neglecting the third which concerns the prime Q. Let us denote this $\mathcal{O}_P \otimes E$-ideal by $[A]_{P,E}$. For these ideals in a Dedekind ring the product formula

$$[AB]_{P,E} = [A]_{P,E} \cdot [B]_{P,E} \qquad (10.17)$$

is well known. In fact, if the notion of Dedekind ring is defined by the property that the ideals $\neq 0$ form a group then (10.17) is part of the definition.

Let $t = t_Q \in E$ be a prime element for Q. If an element $y \in [A]_{P,E}$ is multiplied by a sufficiently height power t^e then the product $t^e y$ will satisfy also the third condition in (10.15) and thus is contained in $[A]_{P,Q}$. Similarly for $[B]_{P,E}$ and for $[AB]_{P,E}$. We conclude: For every element $y \in [AB]_{P,Q}$ there exists e such that

$$t^e y \in [A]_{P,Q} \cdot [B]_{P,Q}. \qquad (10.18)$$

Suppose e is minimal with this property. Our claim is that $e = 0$. We have

$$t^e y = \sum_i a_i b_i$$

with $a_i \in [A]_{P,Q}$, $b_i \in [B]_{P,Q}$ where i ranges over a suitable finite system of indices. Consider the residues $a_i Q \in F$, and choose one among them with minimal v_P-value, say aQ. Then for each i we have $a_i Q = aQ \cdot \alpha_i$ with P-integer elements $\alpha_i \in F$. Thus

$$a_i \equiv a\alpha_i \bmod Q.$$

It follows

$$t^e y \equiv a\left(\sum_i \alpha_i b_i\right) \bmod Q$$

$$\equiv ab \bmod Q.$$

The element $b = \sum_i \alpha_i b_i$ is contained in $[B]_{P,Q}$ since the $\alpha_i \in F$ are P-integers by construction. If $e > 0$ then it would follow $ab \equiv 0 \bmod Q$ and hence either $a \equiv 0 \bmod Q$ or $b \equiv 0 \bmod Q$. In each of these cases division by t would show that (10.18) holds with exponent $e - 1$ instead of e. Then e would not be minimal. This shows that the product formula (10.16) holds.

I claim that for $A = (x)_{tr}$ we have

$$v_Q \mathfrak{D}(A, M) := \min_y \{v_Q(yM)\} \tag{10.19}$$

where y ranges over the elements in $[A]_{P,Q}$.

According to our assumption A decomposes into primes of degree 1 over E. Let M' be one of them and $\mu' : F \to E$ be the corresponding isomorphism. The ideal $[M']_{P,Q}$ is generated by the elements $y = z - z\mu'$ with $z \in \mathcal{O}_P$. Modulo M these are mapped onto $yM = z\mu - z\mu'$ and we see that in this case (10.19) holds for M' in place of A. (Remember the definition (10.4) of the different.) Now the product formula (10.16) shows that (10.19) holds also for A. Of course, it suffices to let y range over a set of generators of the ideal $[A]_{P,Q}$.

We have $A = (x)_{tr}$. On first sight it may seem that the ideal $[A]_{P,Q}$ is the principal ideal generated by x. However after looking more closely into the matter you will realize that in general this is not the case. The $\mathcal{O}_P \otimes \mathcal{O}_Q$–ideal generated by x consists of all elements $y \in \mathcal{F}$ satisfying

$$v_{M'}(y) \geq v_{M'}(x) \quad \text{for each transcendental prime } M' \tag{10.20}$$

$$v_P(y) \geq v_P(x)$$

$$v_Q(y) \geq v_Q(x).$$

Compare this with the description (10.15) of the ideal of multiples of $A = (x)_{tr}$. You will see the difference at the contributions of P and Q. Accordingly the ideal $[A]_{P,Q}$ for $A = (x)_{tr}$ is generated by the element

$$y = t_P^{-v_P(x)} x \, t_Q^{-v_Q(x)}$$

where t_P, t_Q are prime elements for P in F and Q in E respectively. The residue modulo M of this element is

$$yM = (t_P \mu)^{-v_P(x)} \cdot x M \cdot t_Q^{-v_Q(x)} \in E \qquad (10.21)$$

and hence by (10.19)

$$v_Q \mathfrak{D}(A, M) = v_Q(yM).$$

Summation over all Q gives the degree in $E|K$. Since the degree of the principal divisor (xM) in E vanishes we obtain

$$\chi(A, M) = -\deg_{E|K}(B\mu) - \deg_{E|K}(C) \qquad (10.22)$$

where I have put

$$B = \prod_P P^{v_P(x)}, \qquad C = \prod_Q Q^{v_Q(x)}. \qquad (10.23)$$

Here, P ranges over the primes of $F|K$ and Q over the primes of $E|K$. Thus B and C are divisors of $F|K$ and of $E|K$ respectively. In fact, B is the algebraic part of the principal divisor $(x)_{\mathcal{F}|E}$, and similarly C is the algebraic part in $E|K$ of $(x)_{\mathcal{F}|F}$.

In order to prove the Lemma 10.12 we have to identify the two terms on the right in (10.22) with the two terms in the definition of ψ in formula (10.11).

We have $(x)_{tr} = A$. Hence $(x)_{\mathcal{F}|E} = AB$. Since the principal divisor $(x)_{\mathcal{F}|E}$ has degree 0 we conclude

$$\deg_{F|K}(B) = \deg_{\mathcal{F}|E}(B) = -\deg_{\mathcal{F}|E}(A). \qquad (10.24)$$

The isomorphic image $B\mu$ is a divisor of $F\mu$ and has there the same degree as B in $F|K$. In E this degree is multiplied by $[E : F\mu] = d'(M)$, hence

$$\deg_{E|K}(B\mu) = -d(A)d'(M).$$

Similarly we have $(x)_{\mathcal{F}|F} = AC$ and hence

$$\deg_{E|K}(C) = -\deg_{\mathcal{F}|F}(A) = -d'(A) = -d'(A)d(M)$$

since M is of degree 1 over E. Hence indeed, the right side in (10.22) is ψ
(A, M).   □

The symmetric bilinear form $\sigma(A, B)$ on the group of coarse divisor classes is
now fully established. I would like to point out again that the essential idea of the
construction consists in building this form around the notion of different. All other
arguments which I have used are algebraic routine and served only to guarantee the
validity of the Lemma. The particular properties of the different will be used in the
following arguments showing that $\sigma(A, A)$ is positive definite.

10.2   Positivity

In this section I am going to show that

$$\sigma(A, A) > 0 \qquad \text{if } A \not\approx 1 \tag{10.25}$$

for any transcendental divisor A of $\mathcal{F}$. By definition (10.13) this is to be read as the
so-called
Inequality of Castelnuovo-Severi:

$$\chi(A, A') < \psi(A, A') \qquad \text{if } A \not\approx 1 \text{ and } A' \approx A, \tag{10.26}$$

provided A, A' have no common prime.

I shall first prove this inequality in the case when $A = M$ is a transcendental
prime of degree 1 over E. In this case it will turn out that (10.26) is essentially
equivalent to the well known "Riemann-Hurwitz formula" in function fields.
Thereafter I shall treat the case for arbitrary A. That case will be reduced to the
first case by using a discriminant estimate which follows from the Riemann-Roch
Theorem.

10.2.1   Applying the Riemann-Hurwitz Formula

Let M be a transcendental prime divisor of degree 1 over E. As above I denote by
$\mu : F \to E$ the corresponding isomorphism. I claim that

$$\sigma(M, M) = 2gd'(M) \tag{10.27}$$

where $g \geq 1$ is the genus and $d'(M) = [E : E\mu]$ according to (10.10).

In order to find a divisor $M' \approx M$ which does not contain M choose a separating
element $x \in F$ and consider the element $x - x\mu$ in the function field $\mathcal{F}|E$. Its
zeros $M_1, \ldots, M_n$ are those primes whose residue homomorphism sends x onto
$x\mu$. These are transcendental and mutually different from each other. Our M is one

of them, say, $M = M_1$. After suitably enlarging E we may assume that each M_i is of degree 1 over E. (See the remark on page 162.) The pole divisor of $x - x\mu$ in $\mathrm{Div}(\mathcal{F}|E)$ is the same as the pole divisor of x in $F|K$, hence algebraic. Thus

$$(x - x\mu)_{tr} \approx MM_2 \cdots M_n \approx 1$$

and therefore

$$M' := (M_2 \cdots M_n)^{-1} \approx M . \tag{10.28}$$

For each M_i the corresponding isomorphism $\mu_i : F \to E$ coincides with μ on the rational function field $K(x)$. Consequently $\mu_i = \mu\tau_i$ where τ_i is an isomorphism of $F\mu$ into E which leaves $K(x\mu)$ elementwise fixed.

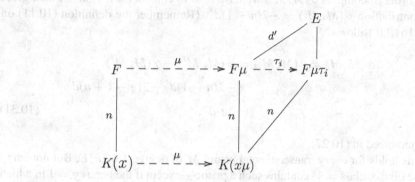

In the above diagram I have written $d' := d'(M) = [E : F\mu]$.

The τ_i are pairwise different and their number is

$$n = [F\mu : K(x\mu)] = [F : K(x)] .$$

We compute: (The minus sign on the right hand side in the following formulas corresponds to the exponent -1 in (10.28).)

$$\chi(M, M') = -\chi(M, M_2 \cdots M_n)$$

$$= -\deg_{E|K} \left(\mathfrak{D}(\mu, \mu\tau_2) \cdots \mathfrak{D}(\mu, \mu\tau_n) \right)$$

$$= -\deg_{E|K} \left(\mathfrak{D}(1, \tau_2) \cdots \mathfrak{D}(1, \tau_n) \right)$$

$$= -\deg_{E|K} \mathfrak{D}(F\mu|K(x\mu))$$

$$= -\deg_{F|K} \mathfrak{D}(F|K(x)) \cdot d' .$$

Here we see the different $\mathfrak{D}(F\mu|K(x\mu))$ of the extension $F\mu|K(x\mu)$ appear, which is the isomorphic image under μ of the different $\mathfrak{D}(F|K(x))$. (Remember formula (10.5) on page 160.) The factor d' in the last line is due to the fact that under the embedding $F\mu \subset E$ the degree is multiplied by the field degree $d' = [E : F\mu]$.

The degree of the different $\mathfrak{D}(F|K(x))$ can be read off from the well known Riemann-Hurwitz formula:

$$2g - 2 = \deg_{F|K} \mathfrak{D}(F|K(x)) - 2n \qquad (10.29)$$

where g is the genus of $F|K$. This formula had been transferred to the algebraic theory of function fields in the paper by F.K. Schmidt [Sch31a] which I have discussed in Chap. 4. It follows

$$\chi(M, M') = -2(g - 1 + n)d'. \qquad (10.30)$$

In order to compute $\sigma(M, M')$ we have also to consider $\psi(M, M')$ which gives the contribution $\psi(M, M') = -2(n - 1)d'$. (Remember the definition (10.11) on page 161.) It follows

$$\begin{aligned}
\sigma(M, M) = \sigma(M, M') &= \psi(M, M') - \chi(M, M') \\
&= -2(n - 1)d' + 2(g - 1 + n)d' \\
&= 2g \cdot d', \qquad (10.31)
\end{aligned}$$

as announced in (10.27).

This holds for every transcendental prime M of degree 1 of $\mathcal{F}|E$. But not every coarse divisor class $\neq 1$) contains such a prime—except if the genus $g = 1$ in which case the proof of (10.25) is complete.

REMARK In the elliptic case, i.e., if $g = 1$, the term $d'(M) = [E : F\mu]$ is what Hasse had called the "norm" $\mathcal{N}(\mu)$. We see that in this case our quadratic form σ is the same as Hasse's $\mathcal{N}$, up to the inessential factor 2. The fact that $d'(M)$ also appears for $g > 1$ confirms Hasse's idea that his "norm" will also play a role for higher genus. (See his letter to Weil of 12 July 1936, cited on page 148.)

10.2.2 The Discriminant Estimate

Let us now deal with the case of genus $g > 1$.

Definition *An integer divisor $A \in \mathcal{F}|E$ is called "non-special" if* $\dim A = 1$ *and A has no multiple prime components.*

This should also exclude those transcendental primes whose residue field is inseparable over E. Hence A remains non-special after arbitrary base field extensions.

I shall have to use the

Lemma *Every coarse divisor class $\not\sim 1$ of $\mathcal{F}|E$ contains a non-special divisor of degree g.*

To verify this, choose a prime divisor Q of $E|K$. As usual we identify Q with its unique extension as a prime of $\mathcal{F}|F$. Its residue field is F, thus Q leads to a good reduction $Q : \mathcal{F}|E \to F|K$. With this goes a divisor homomorphism $A \mapsto \overline{A}$ of $\mathrm{Div}(\mathcal{F}|E)$ to $\mathrm{Div}(F|K)$ which preserves degree and divisor equivalence, and which leaves algebraic divisors fixed. (Compare with what I have said about good reduction on page 145. Thus, although Deuring's creation of a systematic theory of good reduction had to wait until 1942, its essential content was already known to Hasse and Deuring in the year 1937, and it was used by them on several occasions.)

Now, starting from an arbitrary divisor A in the given coarse equivalence class we observe that for any algebraic divisor $B \in \mathrm{Div}(F|K)$ the divisor $A' = A(\overline{A})^{-1}B$ is in the same coarse class. We have $\overline{A'} = B$. Choose B as a non-special divisor of degree g. (It is well known that every function field $F|K$ with algebraically closed base field K contains infinitely many non-special divisors of degree g.) Then A' is of degree g and hence there exists an integer divisor $A'' \sim A'$. We have $\overline{A''} \sim \overline{A'} = B$. Since $\dim(B) = 1$ it follows $\overline{A''} = B$. From the general theory of good reduction we have $\dim A'' \leq \dim \overline{A''} = \dim B = 1$. (See page 125.) Hence $\dim A'' = 1$ and A'' itself is non-special. By construction A'' belongs to the same coarse divisor class as A.

In view of the Lemma we may assume, for the proof of (10.2), that A is non-special of degree g. Let A_{tr} be its transcendental part and A_0 its algebraic part, so that

$$A = A_{tr}A_0.$$

Then $\sigma(A, A) = \sigma(A_{tr}, A_{tr})$. I claim that

$$\sigma(A_{tr}, A_{tr}) \geq 2d'(A_{tr}) \qquad \text{if} \quad A_{tr} \neq 1. \tag{10.32}$$

where $d'(A_{tr})$ is the degree of A_{tr} over F (see the definition on page 161).

After enlarging E if necessary we may assume that all components of A_{tr} are of degree 1. (See the remark on page 162.) We write

$$A_{tr} = M_1 M_2 \cdots M_r \qquad \text{with } r \leq g$$

where the M_i are different transcendental primes of degree 1 with corresponding isomorphisms $\mu_i : F \to E$. We have $\mu_i \neq \mu_j$ for $i \neq j$. Put

$$d_i' = d'(M_i), \qquad d' = \sum_{1 \leq i \leq r} d_i' = d'(A_{tr}). \tag{10.33}$$

We compute using (10.27): (I assume $r > 1$ since the case $r = 1$ has been discussed already and would be trivial in the following discussion.)

$$\sigma(A_{tr}, A_{tr}) = \sum_i \sigma(M_i, M_i) + \sum_{i \neq j} \sigma(M_i, M_j)$$

$$= 2gd' + \sum_{i \neq j} \psi(M_i, M_j) - \sum_{i \neq j} \chi(M_i, M_j)$$

$$= 2gd' + 2(r-1)d' - \chi \left(\prod_{i \neq j} \mathfrak{D}(\mu_i, \mu_j) \right).$$

The divisor $\prod_{i \neq j} \mathfrak{D}(\mu_i, \mu_j)$ is called the *discriminant* of $\mu_1, \ldots, \mu_r$ and is denoted by $\mathfrak{d}(\mu_1, \mu_2, \ldots \mu_r)$, or briefly $\mathfrak{d}(\mu)$. (Here I follow the notation of Hasse in his book [Has02] where capital $\mathfrak{D}(\cdots)$ is used for the different divisor and lowercase $\mathfrak{d}(\cdots)$ for the discriminant divisor. In the classical case, when the $\mu_1, \ldots, \mu_r$ are the isomorphisms of a finite separable extension $F|F_0$ into its algebraic closure then $\mathfrak{d}(\mu)$ coincides with the usual discriminant $\mathfrak{d}(F|F_0)$ of this extension.)

Thus the proof of the inequality (10.32) is reduced to the

Discriminant estimate:

$$\chi \, \mathfrak{d}(\mu) \leq 2(g + r - 2)d' \tag{10.34}$$

which then will show (10.32).

Let Q be a prime of E. The local definition of the different on page 159 yields for the discriminant:

$$v_Q(\mathfrak{d}(\mu)) = \min_x v_Q \left(\prod_{i \neq j} (x\mu_i - x\mu_j) \right) \tag{10.35}$$

where x ranges over those elements in F for which all $x\mu_j$ are Q-integer.

Let us write $P_j = \mu_j Q$ ($1 \leq j \leq r$). These primes are not necessarily distinct. The above condition says that x should be P_j-integral for every P_j.

In the classical case, when discussing the discriminant of a finite separable extension, it is well known that discriminants can also be computed by suitable determinants. The same applies here in our more general case:

First we notice that the product appearing on the right side in (10.35) can be written as the square of a determinant:

$$\prod_{i \neq j} (x\mu_i - x\mu_j) = \pm \det \begin{vmatrix} 1 & x\mu_1 & x^2\mu_1 & \cdots & x^{r-1}\mu_1 \\ 1 & x\mu_2 & x^2\mu_2 & \cdots & x^{r-1}\mu_2 \\ \cdots & \cdots & \cdots & \cdots & \cdots \\ 1 & x\mu_r & x^2\mu_r & \cdots & x^{r-1}\mu_r \end{vmatrix}^2 \tag{10.36}$$

On the other hand, let $u_1, u_2, \ldots, u_r$ be any K-linearly independent elements in F which are P_j-integer for every P_j. Consider their determinant

$$d_\mu(u) := \det \begin{vmatrix} u_1\mu_1 & u_2\mu_1 & \cdots & u_r\mu_1 \\ u_1\mu_2 & u_2\mu_2 & \cdots & u_r\mu_2 \\ \cdots & \cdots & \cdots & \cdots \\ u_1\mu_r & u_2\mu_r & \cdots & u_r\mu_r \end{vmatrix} \tag{10.37}$$

Developing each u_i into a local power series with respect to a common prime element $x \in F$ for the P_j one obtains a Q-adic series for $d_\mu(u)$ having the determinant appearing in (10.36) as its lowest term (with some coefficient in K). The other terms are the determinants which appear when replacing $1, x, x^2, \ldots x^{r-1}$ on the right side of (10.36) with $x^{k_0}, x^{k_1}, \ldots, x^{k_{r-1}}$ where $k_0 < k_1 < \cdots < k_{r-1}$ (again with coefficients in K). Conclusion:

$$v_Q(\eth(\mu)) \leq 2\, v_Q(d_\mu(u)). \tag{10.38}$$

The factor 2 here corresponds to the exponent 2 in (10.36). This leads to the following statement:

$$v_Q(\eth(\mu)) = 2 \min_u v_Q(d_\mu(u)) \tag{10.39}$$

where $u = (u_1, u_2, \ldots u_r)$ range over those r-tuples in F for which all $u_i \mu_j$ are Q-integer. In other words: if we put $P_j = \mu_j Q$ then the u_i should be P_j-integer for $1 \leq i, j \leq r$.

This statement generalizes what Hasse in his book [Has02] has called "Second Discriminant Theorem".

This being said, let B be an integer divisor of degree $g + r - 2$. This is the term appearing on the right side of the discriminant estimate (10.34) which is to be proved. Choose the $u_i \in \mathcal{L}(B)$. Then (10.38) holds for those primes Q of E for which none of the P_j appear in B. There are only finitely many exceptional Q, for which at least of the corresponding P_j appears in B. Choose $b \in F$ such that $v_{P_j}(b) = v_{P_j}(B)$ for each of these exceptions. Then the bu_i are P_j-integer for all these exceptions and hence

$$v_Q(\eth(\mu)) \leq 2\, v_Q(d_\mu(bu)).$$

for those exceptional primes Q. Note that

$$d_\mu(bu) = d_\mu(u) \cdot \prod_i b\mu_i$$

which gives for those exceptional Q:

$$v_Q(\eth(\mu)) \leq 2\, v_Q(d_\mu(u)) + \sum_i v_Q(B\mu_i) \tag{10.40}$$

since $v_Q(b\mu_i) \leq v_Q(B\mu_i)$ by construction. Summing over all Q gives the degree in E. The degree of the principal divisor $(d_\mu(u))$ vanishes. The degree of $B\mu_i$ equals the degree of B multiplied with $d_i' = [E : F\mu_i]$. This gives

$$\chi(\mathfrak{d}(\mu)) \leq 2 \sum_i (g + r - 2)d_i' = 2(g + r - 2)d'$$

which is the discriminant estimate (10.34).

   If this would not be the manuscript for a book but if I would present this proof in a lecture course then I would stop at this point and look expectantly to the audience, awaiting a reaction, perhaps a comment, or a question, or protest. I would hope that someone in the audience would have seen that the proof of the discriminant estimate as given above is not yet complete. The question I would have waited for is:

What if the determinant $d_\mu(u) = 0$?

Well, if this would be the case then the proof would break down. So I have to look for an argument that indeed $d_\mu(u) \neq 0$ for a suitable integer divisor B of degree $g + r - 2$ and suitable elements $u_1, \ldots u_r$ of $\mathcal{L}(B)$. Note that until now I have not yet used the property of A to be non-special. This will now come into the play.

   Let W be a canonical divisor of $F|K$. Its degree is $2g - 2$. Let us choose a divisor B of $F|K$ such that

$$B \sim WA_0^{-1} \qquad\qquad (10.41)$$

where A_0 is, as above, the algebraic part of the non-special divisor A. The degree of A_0 is $g - r$ and hence

$$\deg(B) = g + r - 2$$

as required in (10.34). The Riemann-Roch Theorem shows

$$\dim(B) = r - 1 + \dim(A_0) = r$$

since A_0 is non-special as part of A and hence $\dim(A_0) = 1$. In particular we see that B can be chosen to be integer in its class. I claim that for this divisor B the determinant $d_\mu(u) \neq 0$, for any basis $u_1, \ldots, u_r$ of $\mathcal{L}_{F|K}(B)$.

   Here I have used the index $F|K$ for $\mathcal{L}$ in order to indicate that this module of multiples is to be taken in the function field $F|K$. Similarly $\mathcal{L}_{\mathcal{F}|E}$ refers to the function field $\mathcal{F}|E$. Since

$$\mathcal{L}_{\mathcal{F}|E}(B) = \mathcal{L}_{F|K}(B) \otimes_K E$$

it suffices to exhibit an E-basis $u_1, \ldots, u_r$ of $\mathcal{L}_{\mathcal{F}|E}(B)$ for which $d_\mu(u) \neq 0$. Recall that the maps $\mu_i : F \to E$ are induced by the residue maps of the transcendental primes M_i which appear in A_{tr}. As such they extend uniquely to E-linear maps $F \otimes E \to E$ and hence to maps $\mu_i : \mathcal{L}_{\mathcal{F}|E}(B) \to E$. In this way $d_\mu(u) \in E$ is well defined for any E-basis $u_1 \ldots, u_r$ of $\mathcal{L}_{\mathcal{F}|E}(B)$.

Consider those elements from $\mathcal{L}_{\mathcal{F}|E}(B)$ which are divisible by all but one of the transcendental primes in A_{tr}, say, by $M_2 M_3 \cdots M_r$. They form an E-submodule, namely $\mathcal{L}_{\mathcal{F}|E}(BM_2^{-1} \cdots M_r^{-1}) = \mathcal{L}_{\mathcal{F}|E}(WA^{-1}M_1)$. We have

$$\deg WA^{-1}M_1 = g - 1 .$$

By Riemann-Roch we compute

$$\dim WA^{-1}M_1 = g - 1 - g + 1 + \dim AM_1^{-1} = 1$$

since AM_1^{-1} is non-special as part of the non-special divisor A. Let u_1 be a generator of this submodule. By definition $u_1 \mu_i = 0$ for $i > 1$. But $u_1 \mu_1 \neq 0$ since the module $\mathcal{L}_{\mathcal{F}|E}(WA^{-1})$ is of dimension 0 by Riemann-Roch. After multiplying by a suitable factor from E we may assume that $u_1 \mu_1 = 1$.

Similarly for each i we find $u_i \in \mathcal{L}_{\mathcal{F}|E}(B)$ with

$$u_i \mu_j = \begin{cases} 1 \text{ if } i = j \\ 0 \text{ if } i \neq j \end{cases} \quad \text{for } 1 \leq i, j \leq r .$$

Hence $d_\mu(u)$ is the determinant of the unit matrix.

REMARK A detailed investigation of the above determinant $d_\mu(u)$ shows that in (10.34) and hence in (10.32) we even have equality "$=$" instead of "$\geq$" if $r > 1$.

10.3   The RHp

The foregoing sections refer to a function field $F|K$ with algebraically closed base field K. But the RHp refers to a function field with finite base field. So let $F_q|K_q$ be a function field with finite base field of q elements. (Sometimes people write $\mathbb{F}_q$ to denote the field with q elements but in the present context I prefer K_q in order to reserve the letter F for function fields.) Let K denote the algebraic closure of K_q and $F = F_q K$ the corresponding base field extension. The results of the foregoing section will be applied to $F|K$.

The RHp is concerned with the number N of primes of degree 1 of $F_q|K_q$. Each such prime P extends uniquely to a prime of $F|K$ which likewise is denoted by P. The proper object for singling out these primes among all primes P of $F|K$ is the *Frobenius meromorphism* π of $F|K$. By definition, $\pi : F \to F^q$ is the unique K-isomorphism of F into itself which induces in F_q the operation $x \mapsto x^q$. For an arbitrary prime P of $F|K$ consider the prime

$$\pi P = N_{F|F^q}(P)\cdot\pi^{-1},$$

with the notation I have introduced on page 94 and used since then. We have

$$\pi P = P \quad\Longleftrightarrow\quad F_q P = K_q, \tag{10.42}$$

i.e., if and only if P induces in F_q a prime of degree 1. I have shown this already in Sect. 7.4.4 (see page 103) while discussing the elliptic case, and it is not necessary to repeat this argument here.

Consider the different $\mathfrak{D}(1,\pi)$ as defined on page 159. Here "1" stands for the unit meromorphism which changes nothing. From what I have said on page 159 it follows that $\mathfrak{D}(1,\pi)$ consists precisely of those primes P of $F|K$ for which $\pi P = P$. For each such P choose a prime element $t \in F_q$; then $v_P\,\mathfrak{D}(1,\pi) = v_P(t^q - t) = 1$. (See page 159.) Hence P appears in $\mathfrak{D}(1,\pi)$ with multiplicity 1. We conclude that

$$\deg_{F|K}\mathfrak{D}(1,\pi) = N. \tag{10.43}$$

This puts into evidence that the different divisor $\mathfrak{D}(1,\pi)$ is the proper object to count the number N of primes of degree 1 of $F_q|K_q$.

However, in the foregoing sections we did not talk about meromorphisms but about transcendental primes. The connection between both notions is as follows: Take a function field $E|K$ which is isomorphic to $F|K$. The isomorphism is denoted by $\delta : F \to E$. With this E the "double field" $\mathcal{F} = FE$ is built.

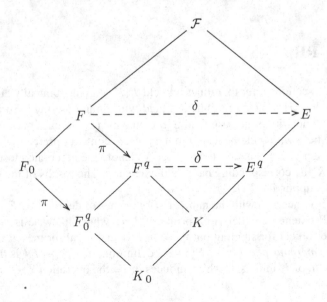

The isomorphism $\delta : F \rightarrow E$ defines a transcendental prime Δ of $\mathcal{F}|E$ with $\deg_{\mathcal{F}|E}(\Delta) = \deg_{\mathcal{F}|F}(\Delta) = 1$. (This divisor Δ can be viewed as a "diagonal", this explains the notation. Compare Sect. 10.5.) Similarly $\pi\delta : F \rightarrow E^q$ defines a transcendental prime Π with $\deg_{\mathcal{F}|E}(\Pi) = 1$ and $\deg_{\mathcal{F}|F}(\Pi) = q$. Hence $\psi(\Delta, \Pi) = 1 + q$. Moreover, the different $\mathfrak{D}(\Delta, \Pi) = \mathfrak{D}(\delta, \pi\delta)$ is just the isomorphic image under δ of $\mathfrak{D}(1, \pi)$ and thus has degree $\chi(\Delta, \Pi) = N$. It follows by definition of σ (see page 161):

$$\sigma(\Delta, \Pi) = 1 + q - N.$$

This being said, consider the subgroup of the group of coarse divisor classes which is generated by the two elements Δ and Π. The positive definite quadratic form σ induces in this subgroup a quadratic form which likewise is positive definite. This implies that the following determinant is positive:

$$\det \begin{vmatrix} \sigma(\Delta, \Delta) & \sigma(\Delta, \Pi) \\ \sigma(\Delta, \Pi) & \sigma(\Pi, \Pi) \end{vmatrix} = \det \begin{vmatrix} 2g & 1+q-N \\ 1+q-N & 2gq \end{vmatrix} = 4g^2q - (1+q-N)^2 \geq 0.$$

Here I have used the value of $\sigma(\Pi, \Pi)$ as determined in (10.2.1) on page 168, and similarly for $\sigma(\Delta, \Delta)$. This gives

$$|N - q - 1| \leq 2g\sqrt{q}, \tag{10.44}$$

which yields the RHp in view of Artin's criterion (see page 52).

10.4 Some Comments

10.4.1 Frobenius Meromorphism

First we observe that this proof uses the *Frobenius meromorphism* π, as did Hasse's proof in the case of genus $g = 1$.

This appears quite natural since π singles out the primes of degree 1 of $F_q|K_q$. Hasse had discovered this in the elliptic case (see page 88). Hasse was quite aware that this fact is not limited to the elliptic case and holds for function fields $F_q|K_q$ of arbitrary genus. But there is a difference. In the elliptic case the primes of degree 1 are interpreted as the rational points of the Jacobian, accordingly their number N appears as the order of the kernel of $\pi - 1$, the latter considered as an endomorphism of the Jacobian J. This had been used by Hasse in writing down $\mathcal{N}(\pi - 1) = N$ (see page 103). In the case of genus $g > 1$ the number N of primes of degree 1 of $F|K$ is in general not the same as the number of rational points of the Jacobian. Therefore, the action of π or $\pi - 1$ on the Jacobian does not play any role in the above proof of the RHp. Instead, N appears as the degree of the different $\mathfrak{D}(1, \pi)$ (see (10.43)).

(Actually, I should have denoted the Frobenius meromorphism by π_q in order to point out that it depends on the base field K_q. In our proof q was fixed, but in order to derive the RHp by Artin's criterion one has to use (10.44) not only for q but also for base fields with order a power of q. See Sect. 4.5. By the way, if $g > 1$ then there exist no other meromorphisms of $F|K$ than π and its powers, (besides automorphisms). This is the reason why the notion of meromorphism is not widely known or used in the theory of function fields. It is of interest in the elliptic case only, due to the fact that for $g = 1$ the function field is at the same time the function field of its Jacobian. And instead of "meromorphism" one now says "isogeny".)

In fact, at no point of the "virtual" proof it was necessary to refer to the action of the coarse divisor classes of $\mathcal{F}|E$ on the Jacobian $J(F)$. Thus Deuring's theorem was not needed at all. (I mean his theorem that the group of coarse divisor classes is isomorphic to the (additive) group of endomorphisms of the Jacobian, see page 145.) The role of Deuring's theorem is reduced to merely provide a motivation to construct the positive definite quadratic form on the group of coarse divisor classes. The Jacobian $J(F)$ and its endomorphism ring did not appear in the proof. Strangely enough this observation has never been explicitly pointed out in the literature. (But implicitly this can be seen in the later paper by Mattuck and Tate who derived the quadratic form from the Riemann-Roch Theorem for surfaces; see Chap. 13.) It suggests that the same or similar arguments of this proof can be used also in other arithmetic problems for function fields, e.g., in fixed point theorems or in questions of rational points of curves over number fields. Some work in this direction has already been done. See, e.g., [Roq76], [Kan80].

10.4.2   Differents

Secondly, we observe that our construction of the quadratic form σ is based on Hilbert's notion of "different" as a divisor. I would like to mention again that Hasse, in his work with elliptic function fields, had already met differents like (10.4) and determined their prime decomposition. (See formula (7.21) on page 101.) Hasse had used it in his proof of the essential norm addition formula. And in his quest for the generalization to function fields of higher genus, he seems to have guessed that again the different would play a role. See his letter to Siegel of 19. December 1935 from which we have cited on page 139. There he wrote that one would have to determine the zeros and poles of a function like $\varphi(\mu z) - \varphi(\nu z)$, given two endomorphisms μ, ν of the Jacobian. But, although Deuring in 1936 had given him a good pass with his theory of correspondences, Hasse did not catch the ball. I have not found any document of those years indicating that Hasse had tried to use differents in the case of genus $g > 1$.

10.4.3 Integer Differentials

In his letter to Weil of 12 July 1936 Hasse mentioned that the "behavior of the differentials of the first kind" under the correspondences was still unknown. (See page 148.) At that time he did not yet know that Deuring had already been able to construct the representation of the group of correspondences M as linear maps of the space of integer differentials[1] of $F|K$ to the space of integer differentials of $E|K$. Deuring informed Hasse about this in his letter of 16. July 1936 but at that date Hasse was already in Oslo at the ICM. Deuring's result appeared in the second part of his paper on correspondences [Deu40]. This representation is not faithful since the space of differentials is of characteristic p while M is of characteristic 0. Certainly, Hasse was well aware of this fact. I do not know what information Hasse expected to gain from the study of this representation. Anyhow, while proving the positive definiteness of σ we have seen that some kind of "period matrix" turned up in the proof, see formula (10.37) on page 171 combined with (10.41) on page 172. The estimate of that determinant was essential for the positive definiteness. Did Hasse have this in mind when he spoke about the abstract analogue of the period matrix?

10.5 The Geometric Language

Geometrically, if $F = K(\Gamma)$ and $E = K(\mathsf{B})$ are interpreted as the function fields of smooth curves Γ, B respectively then the "double field" $\mathcal{F} = FE$ may be interpreted as the field of functions of the surface $\Gamma \times \mathsf{B}$:

$$\mathcal{F} = K(\Gamma \times \mathsf{B}.)$$

All of the above "virtual proof" can be directly translated into the language of algebraic geometry in the following way.

$\Gamma \times \mathsf{B}$ is a surface without singularities. Its prime divisors are the irreducible curves on $\Gamma \times \mathsf{B}$. The local ring of such curve is a valuation ring, which means that the curve is given by a valuation of $\mathcal{F}$. There are three kinds of such valuations which belong to $\Gamma \times \mathsf{B}$: The first two kinds of valuations we have already met in Sect. 10.1 on page 158. These are, firstly, the "*transcendental*" primes which are

---

[1] There is some ambiguity in the use of the words "integral" and "integer". In analysis an "integral" means the result of "integrating" a differential $f(x)dx$, the result being denoted by $\int f(x)dx$ in the notation of Leibniz. In number theory the attribute "integral" is sometimes used in the meaning of "being an integer". In order to avoid misunderstandings I do not use here "integral" in this meaning. Instead, I use "integer" also as an attribute. In the theory of function fields this is interpreted as having no denominator or pole; this can happen not only for divisors but also for differentials. Classically one says "differential of the first kind" instead of "integer differential."

trivial on F and on E, and secondly those primes which are trivial on E but not on F. The latter are called the "*algebraic primes in $F|K$*" and are corresponding to the curves $P \times B$ on $\Gamma \times B$ where P is a point of Γ. But now we have also to consider the "*algebraic primes in $E|K$*" which correspond to the curves $\Gamma \times Q$ with Q a point in B. Thus, if we wish to work on the surface $\Gamma \times B$ then its divisor group is the direct sum of three groups (instead of just two when we considered $\mathcal{F}$ as a function field over E, see page 158):

$$\mathrm{Div}(\Gamma \times B) = \mathrm{Div}_{tr}(\Gamma \times B) \oplus \mathrm{Div}(\Gamma) \oplus \mathrm{Div}(B). \qquad (10.45)$$

I said "direct sum" since in geometry the group operation on divisors is usually written as addition whereas until now I had written it as multiplication in order to point out the analogy to number theory.

An important notion in this geometric environment is that of *intersection multiplicity*. In the general foundation of algebraic geometry the intersection multiplicity of two varieties is not so easy to define. There were quite a number of attempts to obtain a satisfactory algebraic definition in order to include all possible cases. Van der Waerden had been active in this endeavor (in the 1930s) but also a number of other people, including Weil in his "Foundations" [Wei46] since he had not been satisfied with van der Waerden's approach. (Weil had written this in his letter to Artin, cited on page 212.) But in the special situation of curves on a surface of the form $\Gamma \times B$ the situation had been quite clear all the time:

Let M be an irreducible curve (prime) on $\Gamma \times B$. A point (P, Q) of $\Gamma \times B$ is contained in M if and only if its local ring $\mathcal{O}_{P,Q}$ is contained in the valuation ring $\mathcal{O}_M$. Here, $\mathcal{O}_{P,Q}$ is a two-dimensional regular local ring; its maximal ideal $\mathcal{M}_{P,Q}$ is generated by prime elements $t_P \in K(\Gamma)$ and $t_Q \in K(B)$. This ring admits unique prime factorization of elements. M is represented in $\mathcal{O}_{P,Q}$ by a prime element $t_M \in \mathcal{O}_{P,Q}$, and $\mathcal{O}_M$ is the ring of quotients of $\mathcal{O}_{P,Q}$ with respect to the prime ideal generated by t_M. If (P, Q) is also contained in the irreducible curve $N \neq M$ of $\Gamma \times B$ then the intersection multiplicity of (P, Q) in the intersection of M and N is defined as:

$$\dim_K \mathcal{O}_{P,Q}/\langle t_M, t_N \rangle \qquad (10.46)$$

where $\langle t_M, t_N \rangle$ denotes the ideal generated by t_M and t_N. And the global intersection multiplicity $I(M.N)$ is defined as the sum of these numbers over all the (finitely many) points (P, Q) which are contained in M as well as in N. The intersection $M.N$ itself is the 0-cycle of $\Gamma \times B$ consisting of all the above points (P, Q) with the multiplicities given by (10.46).

It is an exercise in commutative algebra to verify that this number $I(M.N)$ equals what I have denoted on page 161 by $\chi(M, N)$ provided $M \neq N$ are transcendental primes.

REMARK The proof of the product formula (10.14) on page 162 could have been somewhat simplified if instead of the tensor product $\mathcal{O}_p \otimes \mathcal{O}_Q$ one would work with the local ring $\mathcal{O}_{P,\,Q}$. For, in this ring the ideals of multiples of divisors become principal. The reason why I have chosen to work with $\mathcal{O}_P \otimes \mathcal{O}_Q$ is that this would have been more in Hasse's style, as he himself has stated in [Has68].

In this way the above "virtual proof" can be entirely translated into the language of algebraic geometry. In fact, what Weil did in 1941 in [Wei41] and also later in the final version [Wei48a] is largely identical with our proof—or better, our proof is largely a translation of Weil's proof. Such a translation had been done in the year 1951 in Oberwolfach by a young student of Hasse. See [Roq53].

What is the conclusion?

As I have said at the beginning of this chapter and shown in the virtual proof, the RHp could have been obtained already in 1937, in the framework of Hasse's theory of function fields on the basis of Deuring's theory of correspondences, since all the necessary prerequisites were available. But now we see that this does also hold with respect to the framework of algebraic geometry. I conclude that the delay in finding the final proof of the RHp was not caused by the fact that the protagonists worked in an improper mathematical framework or language, but by the inherent difficulty to find the way to the envisaged positive definite quadratic form—be it within the theory of function fields or algebraic geometry.

Summary

In this chapter it is put into evidence that the proof of RHp could have been given in 1937 already, following the line which Deuring had proposed. All the prerequisites were available at that time within Hasse's algebraic theory of function fields.

Chapter 11
Intermission

If one would compare our story with a concert, then Artin's thesis together with F.K. Schmidt's paper would pass as the first and second part of the *Introduction* (Chaps. 3 and 4). Hasse's work on the elliptic case (Chap. 7) would be the first movement *allegro assai* with the *theme* set by Davenport (Chap. 6). Deuring's theory of correspondences (Chap. 9) would pass as the second movement *sostenuto*, covering the attempt towards higher genus. The insertion of the virtual proof (Chap. 10) may go as *scherzo allegretto*.

But if we look at the years after 1937 then I could compare those years as *intermission* only, since nothing of importance towards the solution of RHp for genus $g > 1$ can be seen. It was not so that nothing happened towards the RHp, but the activities happened behind the scene (if I am allowed to continue this picture). The musicians practiced their part without coming to the open with a coherent piece. Should I skip this period? I am somewhat hesitating. On the one hand I suppose that at this point the reader will be anxious to come to the *finale presto*, i.e., Chap. 12 describing that and how Hasse's project concerning the RHp was taken over and completed by A. Weil. This would speak in favor of skipping the present Chap. 11. On the other hand, I believe that also small happenings connected to the RHp may be of historic interest. In this spirit I shall now briefly report, in this intermission, on the happenings in the years after Hasse's workshop in 1937, as far as they are connected with the RHp.

11.1 Artin Leaves

Shortly after the Göttingen workshop which ended on Friday 8 January 1937, Hasse visited Artin in Hamburg. I do not know the original purpose of this visit. But it seems not unlikely that he wished to inform Artin about the outcome of the

© Springer Nature Switzerland AG 2018
P. Roquette, *The Riemann Hypothesis in Characteristic p
in Historical Perspective*, Lecture Notes in Mathematics 2222,
https://doi.org/10.1007/978-3-319-99067-5_11

workshop. After all, Artin had been the first who had formulated and investigated the RHp. He had provided an important contribution during Hasse's visit in Hamburg in the year 1932 (see page 57). Surely Hasse remembered also his discussion with Artin in January 1934 when they mapped a road to the RHp for higher genus, which went via the g-dimensional Jacobian variety (see page 133). And now Hasse may have wished to inform Artin about Deuring's concept of correspondences which promised a route circumventing the higher dimensionality of the Jacobian variety (see Sect. 9.3.1). But as said above, I can only speculate why Hasse wished to meet Artin in January 1937.

I have no evidence that at this visit there arose a *mathematical* discussion between the two. But I know that Hasse had heard, either on this visit or already in Göttingen, that Artin had applied for a leave of absence for about one year. Artin intended to follow an invitation as visiting professor to Stanford University in Berkeley, and also from the Carl Schurz Memorial Foundation for a lecture tour through the USA. Artin had submitted his application to the University of Hamburg some days ago, on 5 January 1937.[1]

These news came as a shock to Hasse. For, under the given circumstances this would in all probability lead to the permanent emigration of Artin. His wife Natascha was of half Jewish origin according to the Nazi definitions, and Artin had to worry about the future of his family in Germany. Quite generally, Hasse[2] regarded the forced emigration of so many excellent mathematicians as a heavy loss for science in Germany. In some cases (e.g., for Emmy Noether) he had actively tried to find a way to keep them in Germany but without success. And now Artin too was about to leave. Already in 1934, it seems, there had circulated a rumor that Artin may contemplate to emigrate. I conclude this from a letter of Hasse to Blaschke, dated 31 October 1934, where it reads:

"What you write about Artin I have already heard from Hecke. I am moved a lot by this since it is a shame that we should lose him from Germany."

It is understandable that, when Artin now told him about his decision to accept the invitation to the USA, Hasse tried to talk him into remaining in Hamburg. This was in vain under the circumstances. Hasse's next letter to Artin, dated 14 January 1937, contains heartfelt thanks in particular to Mrs. Artin for the friendly reception in his house, but it contained no mathematics. In the long series of letters between Hasse and Artin this is the first one which does not contain any math.

How Artin was fired from the University of Hamburg and forced to emigrate has been described in detail by Wußing in [Wuß08]. Artin left Hamburg on 20 October 1937. I am citing from a letter of Hasse to Hecke dated 21 October 1937:

---

[1] According to [DS15] it was Lefschetz who had been instrumental in securing the necessary financial means for a stay of Artin in the USA. Recently I have been informed by Karin Reich that, according to Natascha Artin, also Courant had been active in this direction.

[2] And not only Hasse.

"In the meantime you will have managed the farewell parties for the Artins. This affects me really heavy. I have written some goodbye lines to Artin; this was quite difficult for me."

Finally Artin did not go to Stanford but to the University of Notre Dame, Indiana. One year later he changed to Bloomington and 1946 to Princeton University.

During the following years there were only few letters exchanged between Hasse and Artin, and these of personal kind only. See [FLR14]. (After the war the correspondence between Artin and Hasse was revived, and also their friendship). But Artin does not leave our story entirely. We shall meet his name again in the context of Weil's work in Chap. 12.

11.2   The Italian Connection

11.2.1   Severi

As I have reported in Sect. 9.5, Hasse had heard from Weil and from Lefschetz that the solution of his problem, i.e., the algebraic definition of a positive quadratic form on the correspondences, may perhaps be found in the framework of Italian algebraic geometry—provided the geometric statements and proofs can be transferred to characteristic p. Weil had mentioned the "Trattato" of Severi [Sev26]. Therefore, in view of the unsatisfactory outcome of the recent workshop in Göttingen Hasse held a seminar based on Severi's book in the summer semester of 1937. He wished to become more familiar with the concepts and the language of Italian algebraic geometry.

He also wished to get into closer contact to Francesco Severi. The occasion to meet Severi arose soon. The University of Göttingen planned to celebrate its 200th anniversary at the end of June 1937. As usual in academic institutions the celebration was planned with some amount of pomp and, in particular, on an international scale. Delegations of other European and non-European universities were invited. Note that it was the year 1937. Quite a number of foreign academic institutions and academicians refused to participate, as a protest against the expulsion of so many scientists from the German academic scene and more generally against the dictatorial rule of the Nazi government, in particular against antisemitism. Others, who also observed with abhorrence the atrocities of the Nazi regime, decided to participate nevertheless, in order to express to the German colleagues their solidarity in difficult times.

Hasse had heard (from Blaschke) that Francesco Severi, being one of the most prominent academic figures in Italy, also had received an official invitation to the celebration. He took the opportunity and invited Severi to deliver a mathematical lecture. Hasse wrote to him on 19 April 1937:

"...I am taking the liberty to ask you on this occasion to deliver a lecture about your present work. The subject of my seminar in the present semester is based on your book "Trattato di geometria algebraica". The nomenclature of your papers is still somewhat foreign to us but I hope that until the end of the semester my students will be sufficiently acquainted with it."

Severi accepted, and his talk was scheduled for 28 June 1937 at 9 a.m. with the title:

New conceptions and problems of algebraic geometry.

The language of Severi's talk was French.

In another letter dated 17 June 1937 Hasse invited Severi to live in his (Hasse's) house during his stay in Göttingen. Severi gratefully accepted. Thus Hasse and Severi had ample opportunity during the days of the celebration to meet and talk privately, off the formal occasions. Geppert had been asked by Hasse to act as interpreter during this time.

Harald Geppert (1902–1945) had an Italian mother and was fluent in Italian language. He had studied in Breslau and Göttingen and held a *Dozentur* in Giessen 1935–1940, thereafter a position as professor in Berlin. He had done some work in algebraic geometry, in particular on algebraic surfaces. Geppert was well acquainted with the mathematical scene in Italy. Hasse used to ask Geppert for help and advice whenever he wished to establish contacts with Italian mathematicians, and also for translation between German and Italian language. For biographical information about Geppert see [Käh64].

I have not found any record about the content of the conversation between Hasse and Severi in Göttingen. But I believe it is safe to assume that Hasse mentioned his problem concerning the transfer of the "inequality of Castelnuovo-Severi" into the algebraic setting. (See page 166). I am not so sure whether Severi was willing and able to understand the necessity of such transfer.

SIDE REMARK It is an academic tradition that on such jubilee the degree of "honorary doctorate" will be transferred upon some distinguished scientists. Hasse had proposed the name of Severi in this connection and this had been accepted by the Faculty and the University Board of the University of Göttingen. It also had been permitted by the German ministry of education—at that time such permission was absolutely necessary. However, some days before the beginning of the celebration Hasse was asked by the ministry to abstain from the plan of honorary degree for Severi. As the reason for this sudden withdrawal it was said that the German embassy in Rome had not been able to guarantee that Severi does not have "non-Aryan" ancestors. Hasse was much disturbed and sent several letters to people whom he thought to have political influence on the relevant government agencies. Finally, in the last minute the withdrawal was withdrawn and the formal celebration of honorary doctorate for Severi could proceed as planned.

One year later Hasse and Severi met again, at the annual meeting of the German Mathematical Society (DMV) in Baden-Baden. There, Severi delivered a talk on

13 September 1938. In a letter announcing this to Hasse, dated 26 August 1938, he expressed his hope to meet Hasse there. He wrote:

"We could speak about the topic of the note "Application of the theory of algebraic functions" etc. which you sent me and which I have read with interest."

Here Severi refers to Hasse's note [Has37a]. But this note is only part of a collection of somewhat popular essays from the Mathematical-Physical section of the *"Gesellschaft der Wissenschaften zu Göttingen"*. It does not mention any of the special questions and problems which Hasse was confronted with in his project of proving the RHp. It seems to me doubtful whether it could have been a proper basis of a fruitful exchange of scientific views about the problems of algebraization of the theory of correspondences.[3] In any case, during their meeting in Baden-Baden they agreed to the following:

1. Severi would send reprints of his articles dealing with results and methods which possibly could help Hasse in his endeavor.
2. They would try to start an exchange of younger mathematicians from the Italian school of algebraic geometry and the German school of algebra with the aim to establish closer cooperation.

But it turned out that both these schemes did not quite work as planned. As to the first item, I read in a letter of Hasse to Geppert dated 11 November 1938:

"...After my conversation with Sua Eccellenza [Severi] in Baden-Baden I soon received a letter in which he cited almost all papers of Rosati. In particular he mentioned one paper in which the theorem of the minimal equation for correspondences is proved. I have looked up the review about this and other papers of Rosati and Severi in the encyclopedia article by Berzolari. I have not found any rudiment pointing to a purely algebraic treatment of the singular correspondences ... Severi says in his letter that the periods cannot be algebraized but, in his opinion, the topological tools (cicli) could. But I do not see how this can be done ... "

Thus Hasse was still trying to understand the mathematical language of Italian algebraic geometry—apparently without much success. In retrospective we know today that indeed the notion of "cycles" (*cicli*) is a useful tool in algebraic geometry. It seems that Hasse had observed this. But at that time this notion had been introduced into algebraic geometry as a topological tool, and Hasse did not see how to translate it into algebra. Today we know it can be done, as had been foreseen by Severi. In an n-dimensional (non-singular) variety the $(n-1)$-cycles are the

---

[3]On the other hand, this note reveals that Hasse's interest had somewhat shifted, or I should better say expanded, to the theory of function fields over base fields which are algebraic number fields. There are also several other sources which indicate this shift of interest. See, e.g., [Has42a] or the last section of [Has42b].

divisors, and Hasse was used to work with divisors, of course. It appears that he had problems with the notion of lower-dimensional cycles since they are not associated to valuations.

But Deuring in his theory of correspondences had shown how to work, in the case at hand, exclusively with divisors without explicitly using lower-dimensional cycles: In algebraic geometry the "double field" $\mathcal{F} = \mathrm{Quot}(F \otimes E)$ is considered as the function field of a 2-*dimensional* variety, viz. the direct product of two curves. (I am referring to Chap. 10). Among the 1-dimensional cycles of this variety there are what I have called "transcendental" divisors. The intersection of two such transcendental divisors A, B is a 0-cycle. This can be represented by a divisor of the curve with function field E, namely the different $\mathfrak{D}(A, B)$. If Hasse would have followed this idea then he would have to deal exclusively with divisors which he was used to: once with the transcendental divisors of $\mathcal{F}|E$ and thereafter with the divisors of $E|K$. So why did he not follow the path which Deuring had opened, and proceed straight to his envisaged proof of the RHp? Of course we will never know the true reason (if there is one). I have the impression that he did not see the analogy between the notion of intersection of curves on algebraic surfaces, and that of different in algebraic number theory.

As to the envisaged exchange of young mathematicians between Italy and Germany, things could not be realized as planned. From the correspondence Hasse-Geppert it appears that Bompiani was assigned to be in charge of the Italian side for this plan. But after the outbreak of World War II in September 1939 those plans were postponed. Although Hasse on 5 January 1940 still wrote to Bompiani that he wished to have a young Italian mathematician in Göttingen *"in order to learn much of the great results of Italian algebraic geometry"*, he had finally to cancel this. For, on 4 April 1940 Hasse was conscripted to join the research department of the Navy headquarters in Berlin.

11.2.2 *The Volta Congress*

The Royal Academy of Italy in Rome had planned its 9th Volta congress for October 1939. The main theme was set as

Contemporary Mathematics and its Applications.

The organization of this congress seems to have been in the hands of Severi. From Germany a number of mathematicians had been invited, among others I found the names of Blaschke, Caratheodory, Wirtinger, Erhard Schmidt, Süss, Reidemeister, Geppert, and Hasse. This was a relatively large group from Germany. The German ministry named Hasse as the chairman of the delegation. In those years it seemed necessary for political reasons that there was an official chairman in such a situation, the politically correct name in the Nazi jargon was *"Delegationsführer"*. But Hasse did not agree with this nomination. He proposed Blaschke as chairman because Blaschke, he wrote, was much more acquainted with the mathematicians in Italy and their mathematics. However, when Blaschke told him (in a letter of 26 June 1939) there would be the possibility that Bieberbach could be nominated as chairman then Hasse finally accepted. It appears that nobody would have been happy with Bieberbach as *Delegationsführer*.

In July 1939 Hasse started to write down the manuscript for his talk at the Volta congress. The title of the talk was

Rational points on algebraic curves modulo p .

In a letter to Geppert in Gießen dated 18 July 1939 he wrote that he wished to present this manuscript in Italian language, and he asked for help in the translation. Geppert agreed, and Hasse visited him in Gießen for a week in August in order to work with him on the translation. The matter was somewhat urgent since Severi wished to have the manuscripts for the talks by mid-August, so that the printed papers would be available already at the congress in October.

The paper in its printed form has 58 pages. It contains a survey of Hasse's proof of the RHp in the elliptic case, as well as Deuring's algebraic theory of correspondences for function fields of arbitrary genus. As Hasse wrote to Geppert, he considered this paper as a contribution to the mutual understanding of the German and the Italian schools of algebra. But he added:

> "*I hope that the content of this paper will meet the approval of His Excellency [Severi] and will not lead to a disgruntlement towards the German school of algebra. Although my knowledge of the Italian literature is somewhat cursory only, I have the impression that the theory of correspondences, in the form as developed there, cannot be translated directly to the case of a finite base field. For myself, this is the only justification not to be content with a theory based on continuity arguments or transcendental tools (periods, integrals).*"

I have cited this text since here Hasse utters quite clearly his motivation for the search of a purely algebraic base for his proof. Perhaps, if there would have been more time available to become acquainted with the Italian literature, Hasse would have realized that indeed, there did exist a theory of correspondences of algebraic curves, developed in Italy, without essential use of transcencental tools and that his and Deuring's version can be seen as a transfer of geometric ideas into the language of modern algebra.

But time was not available. The Volta Congress which originally had been planned for October 1939, was postponed indefinitely since in September the war (WW II) had begun. Hence Hasse did not get the opportunity to present his view and discuss it with his Italian colleagues. Nevertheless Hasse's manuscript, having been submitted in August already, was printed. I do not know whether and when reprints became available to the mathematical public. The reprint which I found among the Hasse papers carries 1940 as the year of printing. The refereeing journals give 1943 as the year of publication. In the *Jahrbuch der Fortschritte der Mathematik* and *Zentralblatt für Mathematik* the paper was listed but not refereed. After all, in 1943 the RHp had been already proved by Weil and hence large parts of Hasse's paper were outdated. In the *Mathematical Reviews* the referee was O.F.G. Schilling.

The next opportunity to present the algebraic views to an Italian audience arose two years later when Deuring gave a lecture course in Rome at the *"Istituto Nazionale di Alta Matematica"*. It appears that he had been invited by Severi. I do not know whether this invitation was considered to be part of the envisaged exchange between young Italian geometers and young German algebraists. The Deuring lecture was realized from 22 March to 8 April 1941, notwithstanding the difficult war times. There exist lecture notes (in Italian language) [Deu41b]. On 60 pages it contains a brief introduction to the theory of function fields, a report on Hasse's proof of the RHp in the elliptic case, on Deuring's own recent results on the structure of the endomorphism ring in the elliptic case, and on his papers on correspondences.

Thus Deuring lectured about the same topic and with the same aim as Hasse had planned at the cancelled Volta congress. In the lecture notes Deuring does not cite Hasse's manuscript [Has43b]. I do not know whether Deuring had prepared this lecture in accord with Hasse.

There are two footnotes in Deuring's paper which might be of historic interest. The first contains a reference to Weil's Comptes Rendus note of 1940 [Wei40]. There Weil had presented a program how to proceed in the direction of a proof of RHp. I shall discuss that note below in Sect. 12.2. Deuring in his footnote reports that *"by personal communication of the author"* Weil did not have a proof of an important lemma stated in this CR note. Therefore, he said, the RHp still had to be regarded as an open problem. This may be of some interest since several authors claim that already in 1940 Weil was in the possession of a proof. The second footnote was apparently added in the last minute before print. Here Deuring briefly points to a new note by Weil, this time in the Proceedings of the National Academy of Sciences in USA [Wei41]. There Weil announces a proof of RHp. Deuring did not comment this second announcement. Perhaps he had not yet received the final version of the paper.

One year later the 9th Volta congress in Rome was belatedly held (although the war was still going on). The precise dates were 8–12 November 1942. From Germany there participated Blaschke, Caratheodory and Hasse. Thus the German "delegation" was small when compared to the original plans for the congress in 1939. This time Hasse's contribution was not any more confined to the RHp. Instead, as he reported in [Has43a] he spoke generally about *"those methods and results of modern number theory which are based on the arithmetic theory of algebraic functions"*.

The printed version of Hasse's lecture appeared only in the year 1945 after the war [Has45]. The paper is written in German. In the context of this paper the word "arithmetic" refers to those results and methods which are of direct importance to number theory, viz., the theory of algebraic numbers. (Compare this to the former use of "arithmetic" in Artin's thesis which I have discussed on page 17). The paper does not present any new results or methods. It may be regarded as a moment of reflection about the influence of the theory of complex functions on the theory of algebraic numbers. Hasse distinguishes two parts of this influence: "methodical" and "with regard to content". In the first part he mentions Hensel's idea of p-adic numbers and the idea of local versus global in number theory. In the second part he mentions the Mordell-Weil Theorem and Siegel's theorem on integer points of curves. The RHp is mentioned as a side remark only. Hasse refers to his own proof in the elliptic case but also to the newly announced proof of A. Weil in case of curves of arbitrary genus. He says in this respect:

"I have been told that recently A. Weil has succeeded with the proof also in case g > 1. However I have to entertain some doubts in this respect, as long as the envisaged detailed exposition is not available. Two years ago already A. Weil had asserted in a Comptes Rendus note that he was in the possession of a proof, although he had to admit later that he did not have a proof of the decisive fundamental lemma."

Here, Hasse refers to Weil's second note [Wei41].

Perhaps this rather unusual text may be explained by Hasse's disappointment which he experienced when he had heard that Weil, despite his first announcement in [Wei40] did not have a proof at that time.

Hasse in his paper propagates the general idea to investigate the connection between function fields over number fields and function fields over finite fields. In other words: Solution of diophantine equations (global) and diophantine congruences (local). But this is expressed as a vague idea only. The tools for such investigations, i.e., the commutative algebra, were not sufficiently developed at that time.

After reading Hasse's exposition which centers around the mutual influence between number theory and the theory of functions, the reader will wonder why algebraic geometry had not been explicitly mentioned in this context. After all, for some time Hasse had tried to learn about algebraic geometry in order to obtain ideas for the proof of RHp. Has he abandoned this search by now? Perhaps he had done so since Weil seemed to have finally conquered the RHp and Hasse could now turn to other problems.

I have yet to mention an earlier publication of Hasse where he tries to give the algebraic version of the geometry of the Jacobian as g-dimensional variety for $g > 1$ [Has42b]. The manuscript had been submitted in 1941 and appeared in the year 1942. Hasse says that the ideas go back to 1938 when he presented a survey at the annual meeting of the German Mathematical Society about the problems related to the RHp. On that occasion he had met Severi as I have already reported (see page 185). Perhaps this paper can be regarded as Hasse's first step in the direction of adapting the geometric language into algebra, with the aim to finally reach the essential ideas of Italian algebraic geometry as presented in Severi's book—at least those which are of relevance for the study of the Jacobian. But Hasse's paper does not give any substantial contribution. Hasse announces some proofs which he had asked van der Waerden about. The latter answered Hasse's questions but the answer [van47] appeared in the year 1947 only when it was outdated already. Hasse's paper was reviewed by A. Weil in the "Mathematical Reviews". Weil points out that the greater part of the present paper is devoted to an exposition of some of the more elementary properties of the Jacobian variety and adds, that this is

> "... *couched in the arithmetico-algebraic language of the author and his school, which will be familiar to readers of his papers but may act as a deterrent on other classes of readers, and does not seem to the reviewer to be as well adapted to those questions as the language of algebraic geometry.*"

Here we see explicitly stated the reason, as seen by Weil, why Hasse did not succeed with a proof of the RHp for higher genus. Hasse had often propagated that the mathematical notions and hence the language should be adapted to the problems at hand. But the "arithmetico-algebraic" language which Hasse preferred and with which he had obtained all his former outstanding successes was, according to Weil, perhaps not so well adapted in this particular case and prevented him to find his way here. (Already in 1940 Weil had uttered this opinion in a letter to Henri Cartan. See page 205). In his review Weil also points out that the RHp for higher genus has already been proved by himself. For this he refers to his second announcement in 1941 [Wei41]. This fact may explain why Hasse never returned to his attempts to transfer the theory of Jacobians of algebraic curves to characteristic p.

11.3 The French Connection

Let us jump back in time from the 1940s to the 1930s.

In those years Hasse had exchanged mathematical ideas and letters with two French mathematicians: Claude Chevalley and André Weil. The correspondence with Chevalley was about class field theory, while that with Weil was predominantly concerned with function fields and the RHp and hence relevant for our story. I have already mentioned several letters which were exchanged between Hasse and Weil.[4] This correspondence continued in the years after 1936.

11.3.1 On Function Fields

In a letter of 22 November 1937 André Weil wrote to Hasse:

> "*On the occasion of a talk at Julia's seminar I have recently written up a method for establishing the theorem of Riemann-Roch, which is closely related to my proof of the generalized Riemann-Roch Theorem ...*"

Weil refers to his paper [Wei38a] which however had not yet appeared when he wrote this letter. His "generalized Riemann-Roch Theorem" concerns matrix rings

---

[4]See page 134 ff., Sects. 9.4 and 9.5, and page 154.

over function fields in the classical case, i.e., the base field being the field of complex numbers.

Weil continues his letter with an explicit description of his ideas, together with complete details.

On first sight this looks like just another algebraic proof of the Riemann-Roch Theorem which, after all, had already been established by F.K. Schmidt using the method of Dedekind-Weber (see Sect. 4.2). Recently F.K. Schmidt had proved the Riemann-Roch theorem also for those function fields which are not separably generated [Sch36]. Weil explicitly refers to this paper. It seems that he was inspired by it to develop his new approach. Why did Weil find his approach interesting enough not only to mention his setup but sending a detailed exposition in a letter to Hasse? Weil himself gives the answer:

"Thereby a new concept is introduced which in fact coincides with the hitherto used concept of differential . . . "

We see that Weil considered his new concept of "differential" of such basic importance that he immediately informed Hasse about it. Let us see:

Weil introduces the concept of what today is called "adele". An adele α of a function field $F|K$ is given by assigning to each prime P of $F|K$ an element α_P in the completion $\widehat{K}_P$ with the usual finiteness condition, namely that α_P is P-integer for almost all primes P. The function field F is diagonally embedded in its adele ring $\mathcal{A}$, and the Riemann-Roch Theorem appears now as a simple exercise in linear algebra by comparing the arithmetic in F with the arithmetic in $\mathcal{A}$. In this setup Weil's "differentials" appear as K-linear maps $\mathcal{A} \to K$ (continuous in the appropriate topology) which vanish on F. Thus Weil's notion of "differential" of $F|K$ is reduced to the notion of its residues. Weil's condition that it vanishes on F is just the algebraic analogue of Cauchy's theorem, namely that the sum of the local residues of a differential vanishes.

(Here I use "residue" in the sense of the word "residuum" which is used in German; it refers to the local expansion of a differential with respect to a uniformizing parameter t. Then the "residue" of the differential is the coefficient of $t^{-1}dt$; this does not depend on the choice of the uniformizing parameter t).

I have already said in Sect. 8.1 that Hasse had given an algebraic proof of Cauchy's theorem in his paper [Has34d] where he used the classical notion of differential ydx (with $x, y \in F$). But in Weil's setup Cauchy's theorem is taken as an *axiom* for his "differentials" which appear now in some sense as "dual" to the "functions" of the field F—as it was well understood in the classical case of analytic functions over the complex base field $\mathbb{C}$.

The importance of Weil's approach lies in the use of the new notion of "adele ring" which seems to be the proper structure when it comes to study the connection between local and global. Actually, Weil in his letter did not yet use the word "adele"; it seems that he himself created this word at a later occasion. (In any case, the word "adele" appears in [Wei59]). Weil created his adeles as the additive version of Chevalley's multiplicative "idele". Chevalley's idele group is contained in Weil's adele ring as the group of its invertible elements.

Weil could assume that Hasse was familiar with Chevalley's theory of ideles which is expanded in the paper [Che36] of the year 1936. In fact, ideles appear already in a letter of Chevalley to Hasse dated 20 June 1935. Again, Chevalley did not yet use the word "idele" in his letter, he says "élements idéaux" instead. The word "idele" was proposed by Hasse in his enthusiastic *Jahrbuch* review of Chevalley's paper. Chevalley himself used it later in his article on class field theory [Che40]. (By the way, in that later paper Chevalley used the word "differential" in the same way as Weil in his letter to Hasse, but this time in the multiplicative sense—obviously inspired by Weil).

Weil closed the description of his results with the words:

"If you find that the above lines are suitable for publication in Crelle's Journal then they are at your disposal."

Hasse immediately realized the importance of Weil's approach. He wrote to Weil on 30 November 1937:

"Many thanks for your very interesting letter with the beautiful and new proof of the theorem of Riemann-Roch. In particular I like your nice and original idea of introducing differentials. It will be a pleasure for me to accept your kind offer and compose an article for Crelle's Journal from your letter."

Weil's paper appeared 1938 in volume 179 with the subtitle *"Aus einem Brief an H.Hasse"* [Wei38b].

REMARK 11.1 The publication of Weil's article in Crelle's Journal carried a high risk for Hasse if it would have come to the knowledge of higher Nazi officials. Already the exchange of letters with Weil was not without risk. It appears that now Hasse felt his position quite strong in his dealings with the German ministry of education, so that he was prepared to take this risk. By the way, in the preceding year Hasse had already published in Crelle's Journal another paper by a Jewish author, namely by Kurt Hensel [Hen37]. In a similar way this risk existed also in view of his contacts with other people who were outlawed by the Nazis for racial or political reasons, for instance with authors of the new edition of '*Enzyklopädie der Mathematischen Wissenschaften*", whose section for Algebra and Number Theory was edited by Hasse jointly with Hecke. In the latter case, however, the contributions of those authors were finally not allowed to be published, against the proposals of Hasse and Hecke. (Details can be found in the correspondence file Hasse-Hecke which is available at the sources mentioned in the Preface).

REMARK 11.2 Sometimes in the literature it is said that the notion of "adele" had been introduced by Artin in his paper [AW45] under the name of "valuation vector" which later were renamed as "adele" by A. Weil. Here we see that the introduction of adeles is due to Weil himself, in the year 1938 already. However both Weil and Artin make it clear that they were inspired by Chevalley's ideles.

REMARK 11.3 It seems not widely known that Chevalley had discovered his notion of idele and its use in class field theory while working on the text of a survey article for the new edition of *"Enzyklopädie der Mathematischen Wissenschaften"*. In this project Hasse had proposed to Chevalley to write an article on class field theory, and Chevalley had accepted. His letter of 20 June 1935 to Hasse, where he explained his new notion of *"élements ideaux"* and its use in class field theory is a preliminary report about the progress of writing this article. (I have mentioned this letter on page 192 above already). Chevalley did complete his article and he reported about it at the annual meeting of the German Mathematical Society, on 12 September 1938 in Baden-Baden. The article was translated into German by Martin Eichler. But its publication was delayed due to problems with printing during World War II. I know that the manuscript still existed in the year 1948, when Hasse inquired whether Chevalley still wished this manuscript to be published. At that time Chevalley held a position at Columbia University, New York. Apparently he did not wish publication any more. After all, his final paper on the foundation of class field theory with the notion of "idele" had appeared in 1940 already in the *Annals of Mathematics* [Che40]. Anyhow, his original manuscript for the German encyclopedia would have been an important historical document, but it seems to be lost.

It appears that Weil had developed his new concept of differential in the hope that it may be useful for a proof of RHp. For, in a letter of 20 January 1939 Weil wrote:

> *"Recently I have interrupted my investigations on p-groups in order to pick up again some former thoughts about the analogy between number fields and function fields. A first consequence was that I proved the functional equation for the zeta functions with arbitrary characters . . . "*

When Weil speaks of "zeta functions with arbitrary characters" he means what usually is called L-series $L(s \mid \chi)$ for ray class characters χ of a function field over a finite base field. For these functions, a proof of the functional equation required a generalization of the Riemann-Roch Theorem for ray classes. Weil offers to send a note about his proof to the *Göttinger Nachrichten* for publication. But Hasse had to remind him that Witt had proved the functional equation already and that he (Hasse) had informed Weil about Witt's proof in his letter of 12 August 1936. (See page 147). Witt's proof had never been published but in the meantime, Hasse wrote, there had appeared a paper by Weissinger (a student of Artin) containing a proof of the functional equation [Wei37]. Weil thanks in a letter of 9 February 1939. Now he had read Weissinger's paper and found that his own proof was exactly the same as Weissinger's.

In the same letter Weil utters a new idea for the proof of RHp:

> *"The functional equation yields the RHp for those L-series which decompose into L-series of degree 1. If I am not mistaken these are the L-series of those function fields for which you have proved the RHp jointly with Davenport, and this seems to be the actual reason why your proof succeeds. But I suspect that all linear factors of the zeta functions in function fields may be viewed as non-abelian L-series in the sense of Artin. If so then there would follow a proof of RHp. "*

In his reply Hasse confirms Weil's assumption that the function fields of his paper with Davenport are those for which the zeta function decomposes into L-series of degree 1. He adds that cyclic fields with degree a higher power of p have this property as well. (Recall that the Davenport-Hasse fields are certain cyclic extensions of degree p of a rational field; see Sect. 6.3.3). This addition, however, is not true. The L-series of such fields can be treated by using Witt vectors. H.L. Schmid has computed the L-series of those fields in detail [Sch41]. From this it follows that Hasse's statement does not hold for cyclic extensions of degree $p^n > p$.

Weil's idea to regard the linear factors of L-series as non-abelian L-series in the sense of Artin did apparently not work. Weil's later proof of the RHp starts with Deuring's approach as Hasse had originally envisaged. See Sect. 12.3.

The above citations from letters show that there was a lively and prolific exchange of ideas and facts between Hasse and Weil in those years. In fact this was not restricted to the problem of RHp but covered a wider range, in particular function fields (of one or several variables) over number fields.[5]

In a letter of 19 May 1938 Hasse writes that in his seminar they are working towards an algebraization of Weil's thesis [Wei29]. (If I say "they" here then I have in mind Hasse and Siegel. The latter had in the meantime changed from Frankfurt, where he had troubles of political kind, to Göttingen where Hasse tried to protect him from such troubles). Hasse reports that they had been partially successful, in particular with class field theory for abelian unramified extensions of function fields over number fields. In his next letter of 6 January 1939 Hasse writes that in his seminar they are also working on Siegel's great paper [Sie29]. We see that Hasse's interest now includes function fields over number fields as base fields, as I have mentioned earlier already.

Of particular interest seems to be the following query of Weil in his letter of 24 February 1939:

"Have you ever thought whether the higher dimensional varieties over a finite field also have zeta functions?"

Hasse replied on 7 March 1939 that he has never thought about zeta functions of higher dimensional varieties over a finite field but, he adds,

"I did consider the question about zeta functions of one-dimensional varieties over a number field. This would be somewhat similar as for 2-dimensional varieties over a finite field. I have the impression that the very elegant results of Hecke and Petersson point in this direction. I have asked my present student Humbert (Lausanne) to investigate the relationship between Hecke's result and my theory of elliptic function fields."

---

[5]In some letters Weil also touched personal matters. He asked Hasse for recommendation letters when he applied for academic positions in France. However these applications were not successful at that time. He put his frustration into words in a letter to Hasse: *"In France, appointments have little to do with scientific achievements …"*

But Humbert did not stay long in Göttingen. Because of the precarious political situation in Germany 1939 (shortly before the outbreak of the war) he left Göttingen and returned to Switzerland. Thus again Hasse lost a promising student, as he had experienced one year earlier with Hanna von Caemmerer (see page 154).

From Hasse's correspondence with Petersson it is seen that Hasse himself was deeply involved with the problem of the definition and the properties of zeta functions of elliptic curves over number fields. He tried to define such zeta function as product of the zeta functions of the reduced curve modulo p for all prime numbers p. Well, for "almost all" p, namely those for which the curve admits good reduction. The situation was similar to when Artin had defined his new L-series for Galois characters, which originally could be done with the contributions of the unramified primes only. Hasse tried to extract from Petersson's work that his new L-series satisfy a functional equation which then perhaps could lead the way to an adequate definition of the zeta function of an elliptic curve over a number field, accounting also for the primes with bad reduction. However Hasse had to postpone this plan until after the war.

After the war André Weil took over and investigated zeta functions of curves over number fields [Wei52]. However he did not consider elliptic curves but the generalized Fermat curves from the Davenport-Hasse paper (see Sect. 6.3). Weil cited Hasse as the one who originally had the idea. Weil also mentions the name of Humbert in his comments to his paper. See vol. II of [Wei79a]. But Hasse, after all those difficult years, had to admit that he had forgotten about it. Nevertheless Hasse himself now wrote a follow up paper where he expanded his own ideas about this [Has55]. The case of elliptic curves was treated in several papers by Deuring in the 1950s.

11.3.2 The Book

Usually when Hasse had received a paper for publication in Crelle's Journal he himself carefully checked the manuscript and prepared it for publication (see [Roh64]). However, in the case of Weil's paper on the Riemann-Roch Theorem which I have discussed in the preceding section, he did not do this himself but he asked his assistant Ernst Witt to do it. The reason was that Hasse used most of his working time to complete his long projected book on algebraic number fields based on the local theory of p-adic fields in the sense of Hensel. In a letter of 30 November 1937 he had written to Weil:

> "At present I am working hard on my book about number theory which has been projected a long time ago. Volume 1 is completed up to a few paragraphs. It contains a brief presentation on elementary number theory from an advanced view point, an elaborate and quite modern foundation of valuation theory and its application to the foundation of the arithmetic in algebraic number fields, including functions fields... I hope that volume 1 can appear next year."

Several years ago Hasse had signed a contract with Springer for his book. There were to be two volumes, the first as described by Hasse in his letter, and the second volume about class field theory in its new formulation with Artin's reciprocity law and Hasse's Local-Global Principle. But the agreed deadline for submission of the first volume had already been exceeded. (In the letter to F.K. Schmidt of 25 June 1936 Hasse had mentioned the date of 1 November 1936. I have already cited from that letter on page 140). Therefore Springer, through F.K. Schmidt as the managing editor of the so-called yellow series, tried to excert some pressure upon Hasse to finally finish the manuscript, in particular since Springer had given some advancement by financing an assistant who was to help Hasse in this project. Hence for several months now Hasse had put away most of his other work and concentrated on finishing the book. In a letter he writes that he uses 75% of his working time for the book project.

The next letter from Weil to Hasse seems to be lost. It is not contained in Hasse's *Nachlass* since Hasse had forwarded it to Witt who took care of the editing and proof reading for Weil's paper. But from Hasse's reply it can be concluded that the lost letter contained, among other things, an offer of Weil that he would take care of a French translation of Hasse's book. Hasse wrote to Weil on 16 December 1937:

"Many thanks for your kind offer to take care of a French edition of my book. In principle I do agree but I have to consider the wish of my publisher Springer that the French edition should not appear at the same time as the German one, but only $1\frac{1}{2}$ years later at the earliest … In any case there will still be some time for this."

Apparently Weil had not known that already some months earlier Chevalley had asked Hasse (in a letter of 7 July 1937) about the possibility of a French edition, and Hasse had replied to him the same as he now did to Weil. By the way, the appearance of Hasse's book was also expected in England, for Davenport wrote to Hasse on 30 October 1937:

"The appearance of vol. 1 of your book is looked forward to with great interest by all English mathematical circles."

The great interest for Hasse's book appears to have been based on three expectations: First, the book was to contain a systematic treatment of algebraic number theory by means of Hensel's p-adic numbers, based on valuation theory. Second, it included the theory of function fields over finite base fields. In today's terminology, it was a treatment of global fields based on the theory of local fields. This was quite new at that time and not yet treated systematically in a book. Moreover, one hoped that the book contained the basic tools which are helpful to build a proof of the RHp. In addition, since Hasse had planned a second volume presenting class field theory in its modern form, it was expected that he would develop all necessary prerequisites for class field theory in his volume 1.

But alas!, although the first volume was completed in November 1938 it appeared much later than expected (in 1949 only) and the second volume was never written. The story is as follows:

It was on 4 November 1938 when Hasse had finished his book manuscript and posted it to Springer. One day later he wrote to Davenport:

"I feel extremely relieved to have this nightmare off my mind."

But it turned out that the nightmare persisted. F.K. Schmidt wrote on 18 November 1938 to Hasse that the publisher had estimated the size of the book to about 600 pages. This was too much and he proposed to skip some of the sections. Hasse replied on 24 November that he does not wish to do this. In particular he did not want to skip those topics which had not been systematically treated in the literature (e.g., exponential, logarithmic and power function in p-adic fields). Hasse proposed to leave the text as it is.

In a later letter he suggested that Springer could publish this text in 2 volumes and, accordingly, Hasse would be freed from the writing of another volume on class field theory as was originally envisaged. Hasse proposed the name of Chevalley as a possible author of a book on class field theory. But again this was not accepted by Springer. There followed an exchange of several letters but both sides, the author and the publisher, did not give way until finally, in a letter of 9 December 1938, F.K. Schmidt wrote that the decision of this dispute could be postponed since this was not an urgent matter.

In a letter of 6 January 1939 Hasse informed Weil about this state of affairs since Weil had shown interest for a French translation of Weil's book, as we have seen above. Hasse wrote:

"The manuscript of the first volume of my Number Theory is complete. It was to appear in May of this year. But now the publisher had made some troubles since I have exceeded the envisaged size. I was asked to skip from the manuscript what in my opinion are the best pieces. Since I have refused to do so, the publisher has put off the publication for the time. Hence I do not know when the book will appear and whether it will appear at all."

Weil replied on 20 January 1939[6]:

"I am very appalled to hear that the publication of your great Number Theory has been postponed indefinitely."

And Weil adds:

"Under these circumstances, would you not reconsider our former translation plans? On the other hand there may be a possibility here to publish your book in German with Hermann. In this case it would be necessary to cut the book into smaller pieces of 100 or 120 pages, according to the general policy of the "Actualités scientifiques". This would perhaps not be objectionable; but would you like that your book appears in a foreign country? I any case I am prepared to do what is possible in this respect."

---

[6]In the original we read the date "20.I.38" but the content of the letter shows clearly that this was a misprint and the letter was written in 1939.

"Hermann" was the publishing company in France of the series "Actualités scientifiques".

In his reply dated 4 February 1939 Hasse thanks Weil for his interest and engagement for the book. But he would rather like the book first be published in Germany if published at all.

Let me briefly report on the further fate of this book. After some time Hasse and the editors of the "yellow series" came to an agreement: they would ask an independent authority for his opinion. For this service Siegel was chosen, who was willing to do the job. Siegel's recommendation turned out to be clear: Hasse's manuscript should be published as it is. But this is still not the end of the story. In a letter to Ferdinand Springer, F.K. Schmidt expressed his doubts whether Siegel had read the manuscript at all[7] and hence a second reviewer should be asked. But Springer categorically decided that this is now the end of the story and Hasse's book should appear as it is.[8] Obviously Springer did not wish to have further disagreements with both Hasse and Siegel. And so Hasse's manuscript was accepted for publication.

This happened in the summer of 1939. As is well known, on 1 September 1939 Hitler started what was to become the second world war. During war time the supply of paper was scarce and publishing activities were heavily restricted. This applied also to Hasse's book. After the war things slowly normalized. Hasse had left Göttingen and gone to Berlin where he was a member of the Academy of Sciences. His book finally appeared in the year 1949 but in the publishing house of the Academy, not in Springer Verlag. Hasse had made a deal with Springer that he would write another book on Number Theory, containing his notes from his lecture courses in Berlin, and on the other side Springer would admit that the former manuscript would be published by the Berlin Academy. Hence there were two books by Hasse on Number Theory: the "yellow Hasse" (because the Springer books from the "*Grundlehren*" series had a yellow cover) and the "blue Hasse" (the color of the cover of the Academy books). The blue Hasse became a classic, it had several editions, finally it became yellow (i.e., it was taken over by Springer Verlag) and was translated into English, the last edition was in the year 2002.

11.3.3   Paris and Strassbourg

In a letter of Weil to Hasse dated 19 February 1938 he wrote from Strassbourg (where Weil held a position):

"Since you will come to Paris soon, I would be particularly glad if I could be sure to meet you there. Please let me know in advance when you come, for I

---

[7]In view of all what is known of stories about Siegel this may well have been the case.

[8]I have found these letters in the archive of Springer Verlag Heidelberg.

do not travel to Paris as often as in earlier times. It would hardly be a detour if you would visit Strassbourg either on your trip to Paris or on your return. I very much hope that you may be persuaded to do this, what do you think?"

It appears that Weil had obtained some information that an invitation of Hasse to Paris was under way. Hasse received the official invitation several days later, on 27 February 1938. (It may be that the official invitation had been delayed through censorship in Germany). The invitation was signed by the Dean of the *Faculté des Sciences* and included three scientific lectures at the Sorbonne. It was Gaston Julia who had recommended this to his Dean. Julia had met Hasse in Göttingen on the occasion of the University centennial in June 1937, and thereafter they had exchanged some letters but not of mathematical content since their mathematical interests were quite different.

The invitation to Paris was timed for May 1938. But this timing was not convenient for Hasse. As he wrote to Julia on 28 February 1938:

"Before officially replying to your dean I would like to describe to you personally my actual situation. Presently I am engaged with all my power writing up my book on number theory. The book had been planned now for 10 years, and the publisher is demanding its completion with unusual intensity. Hence I am unable to direct my attention towards any other topic before this book is completed. I am estimating this to be not before June."

Again we see that the writing of the book forced Hasse to put off all other work. (For the same reason he also canceled a visit with Davenport which had been planned for the summer of 1938). Accordingly he proposed to shift his visit to Paris from the summer to the beginning of the winter semester. But several days later he sent another letter to Julia, dated 7 March 1938:

"I am asking myself whether it would be better to move the whole thing until next spring. For, Chevalley told me that he is visiting the USA for half a year, starting on the first of October. It would be a pity if I would miss him in Paris."

This suggests that it had been Chevalley who originally had proposed to Julia to invite Hasse to Paris. During the past years there had been continued cooperation between Hasse and Chevalley. In the summer of 1933 Chevalley had stayed in Marburg with Hasse and from then on they had exchanged letters, mostly on class field theory. Since Chevalley and other young founders of Bourbaki sometimes met in Julia's seminar it seems not improbable that Julia's invitation of Hasse to Paris had been suggested by Chevalley.[9]

Hasse had replied to Weil's letter on 24 February 1938. He informed Weil that his visit to Paris will be postponed. But he added:

---

[9]Later however, in a letter of 20 January 1939, Weil writes: "We have had heavy differences with Julia ..." (*"Mit Julia haben wir uns verkracht..."*).

"If this [the visit to Paris] *will be realized then it will be a great pleasure to me to travel via the much beloved Strasbourg and to visit you there."*

Hasse's visit to Paris was finally set for May 1939. But later it turned out that this timing was quite unfavorable for Hasse. On 20 January 1939 Weil wrote:

"...It's a great pity that you could accept the invitation to Paris in this year only. For, last year all our number theorists would have been there. But this year in May, by an unfortunate coincidence they will all be out of town (if not Chevalley will be back from America which may be possible). Pisot will be in Göttingen, [10] *Chabauty in Manchester with Mordell, myself in Cambridge since I just got a stipend for study in England and Scandinavia. I don't have to say how much I regret to have to miss the opportunity to meet you after such a long time ..."*

As it turned out, Chevalley remained in Princeton, thus Hasse indeed did not meet any of the mentioned number theorists on his visit to Paris.

A few days before Hasse's start for Paris he received a letter from Elie Cartan saying that:

"My friend M. Julia is very weak and has been on leave for more than three months. Hence I myself have to determine today the days and time which are suitable for your three lectures at the Institute Poincaré ..."

These lectures were finally set for 19, 23 and 24 May 1939. But Hasse arrived already some days earlier in order to participate at the ceremonies of the Paris Academy for E. Cartan's 70th birthday, which took place on 18 May. (On this occasion Hasse again met Severi). There in a short address he conveyed to E. Cartan the greetings and wishes of the DMV (*Deutsche Mathematiker Vereinigung*). I did not find the text of this address in the *Nachlass* of Hasse. But from Cartan's letter of thanks some days later it may perhaps be deduced what Hasse had said. E. Cartan wrote:

"...I am glad that my papers and my books have found interest among the German mathematicians and that I can count in Germany with a great number of friends and scholars."

But then he added:

"It is among my heartiest wishes that the close international collaboration will be maintained against all obstacles; it is impossible that the people in the world of science and art, when all passions are excluded, do not consider themselves as friendly partners (ne sentent frères)."

---

[10] *Charles Pisot (1910–1984) had a research grant to study number theory in Göttingen in the summer semester 1939. He was born in Alsace hence he spoke German as well as French. As Weil wrote to Hasse it was Pisot which he had had in mind for translation of Hasse's book into French. Pisot became a member of Bourbaki for some time.*

Perhaps this can be read as kind of critical reminder of certain tendencies propagated by some German mathematicians like Bieberbach et al., to establish what was named *"Deutsche Mathematik"* in contrast to mathematics from other countries or cultures. By the way, Hasse and his students (except Teichmüller) never published a paper in the journal *"Deutsche Mathematik"* which had been established by Bieberbach.

Hasse's three lectures at the Sorbonne were announced with the title:

New investigations on the arithmetic of algebraic function fields.

I. Generalities: Divisor class group and ring of multipliers.
II. Rational and integer points on algebraic curves with integer coefficients.
III. Rational points on algebraic curves with coefficients mod. p .

Hasse's lectures were read in French. He had prepared his manuscript with the help of a language teacher.

Despite Julia's health problems he did attend Hasse's lectures.[11] Unfortunately I did not find the text of these lectures. From the correspondence Hasse-Julia it appears that Hasse was to send the manuscript to F. Roger who was supposed to have it mimeographed. At that time Roger was the secretary of Julia's seminar. In any case the manuscript seems to be lost during the following turbulent times.

On the way home from Paris Hasse went through Strasbourg following the invitation by A. Weil last year (see page 198). However, just at this time Weil was away on a trip to Britain and Scandinavia, as already mentioned above. Instead, Hasse was friendly received by Henri Cartan and Ehresmann, as Hasse later reported to Julia. He gave a colloquium talk with a small audience (*"in kleinem Kreis"*). It is said that he repeated his third lecture which he had delivered in Paris. H. Cartan wrote notes of Hasse's lecture into his personal diary. I am indebted to Professor Michèle Audin for sending me copies of the relevant pages of H. Cartan's diary which is preserved. From those notes it can be seen that Hasse in Strasbourg just presented his proof of the RHp in the elliptic case, and over the base field $\mathbb{F}_p$ only. At the end of Cartan's notes we find the sentence that this can be generalized over an arbitrary finite base field, and for curves of arbitrary genus. But Deuring's paper for higher genus is not mentioned.

Thus again, as it was with Hasse's Italian lectures in Rome, his lectures in Paris (and Strasbourg) did not contain anything new in the direction of the RHp, and they had no visible effect on future research.

---

[11] Gaston Julia had been seriously injured during World War I and since then had health problems throughout his life. Nevertheless, and perhaps just because of this experience, he vehemently advocated a close political cooperation between France and Germany, so that there would be no further war between the two countries ...

Summary

In the years after the Göttingen workshop in 1937 there was no essential progress in the direction of the RHp for function fields of higher genus. I have reported in this chapter about the following topics:

1. Artin was forced to emigrate to USA due to the Nazi legislature. Thus Hasse lost a friend and a valuable partner for mathematical discussion.
2. Since Weil had suggested that the solution of the problem may be extracted from Italian algebraic geometry, Hasse tried to familiarize himself with algebraic geometry. In particular he met Severi several times, but without much avail.
3. Hasse kept his contact to Weil by exchanging letters. This led to a publication of Weil on function fields in Crelle's Journal, and also to an exchange of some interesting ideas on zeta functions. But again, concrete steps towards a proof of the RHp did not appear. A planned meeting of Hasse and Weil in Paris could not be realized.

Chapter 12
A. Weil

Paris, 1929

André Weil (1906–1998) was 8 years younger than Hasse. He was born and raised in Paris. He received his doctorate 1928 at the University of Paris, supervised by Hadamard, with his thesis "*Arithmetic of algebraic curves*" where he proved his part of what today is called the Mordell-Weil Theorem. His name appeared already

© Springer Nature Switzerland AG 2018
P. Roquette, *The Riemann Hypothesis in Characteristic* p
in Historical Perspective, Lecture Notes in Mathematics 2222,
https://doi.org/10.1007/978-3-319-99067-5_12

several times in our story since he had exchanged letters with Hasse and had early shown interest in the RHp.

12.1   Bonne Nouvelle

At the outbreak of World War II our story took a sharp turn.[1] Hasse was conscripted to work at the Navy headquarters in Berlin. There he could not do much research work in the direction of the RHp, as he himself stated on several occasions. A. Weil who stayed in Finland when war was declared, was suspected to be a Soviet spy and he was arrested. (Finland was in the state of war with the Soviet Union 1939–1940 about Karelia.) Apparently it was Nevanlinna and/or Ahlfors who was finally able to arrange that he could return to France. But there again Weil found himself arrested, as a conscientious objector to military service. He was imprisoned in Rouen, the prison building being called "Bonne Nouvelle". According to his own words, while waiting for the trial he found ample time to work in mathematics (see [Wei79a]). He had the chance to get into contact with his family and his friends and colleagues. In particular the exchange of letters with his friend Henri Cartan continued during this period. The highly interesting correspondence between H. Cartan and A. Weil is edited by Michèle Audin [Aud11]. Reading those letters gives a lively picture of the various mathematical ideas which Weil fostered in this period. Among them the RHp occupies a prominent place. The following citations I have taken from Audin's book, and translated into English.

On 26 March 1940 Weil wrote from Bonne Nouvelle in a letter to Cartan in Strasbourg:

"... *My arithmetic-algebraic investigations are gaining momentum (I believe I have obtained very important results about the ζ function of algebraic function fields). I do urgently need that you answer my former question as early as possible: In an algebraic function field (field of constants of characteristic p) the number of classes with n-th power being 1. If you find the result for genus 1 do not be content with it but continue the research until you find the result for arbitrary genus, whether it is published (in Crelle's Journal.)*"

Weil refers to a former letter where he had asked Cartan to find out whether this number (h_n in the notation of Hasse) had been determined already by Hasse or by any of his students. In his answer of 2 April 1940 Cartan wrote that he had been unable to do the search which Weil had asked for, because the relevant volumes of Crelle's Journal were still at the bookbindery at Strasbourg and not yet returned. But

---

[1]And not only our story.

he had checked the volumes of Zentralblatt, at least until 1937—the rest was also at the bookbindery. Here is what he found:

1. Hasse's paper [Has36c] where he determined the number h_n for elliptic function fields. (See formula (8.13) on page 113.)
2. The paper of Hasse-Witt [HW36] where h_n is determined for a p-power $n = p^r$, this time for function fields of arbitrary genus. (See Sect. 8.2.1, and also the remark on page 115.)
3. The paper of Deuring [Deu36] which contains a partial result for $n \not\equiv 0 \bmod p$. In general it was conjectured that $h_n = n^{2g}$ for $n \not\equiv 0 \bmod p$. Deuring shows that if $K|K_0$ is cyclic of prime degree and if the conjecture holds for K_0 then it holds for K.

At the end of Deuring's paper there is a remark "added in proof" which says:

"The solvability of $X^n = C$ for every class C of degree 0 and every exponent n can be easily proved by Hasse's method."

In this connection "solvability" seems to include the determination of the number of solutions. When Deuring mentions "Hasse's method" then he probably means the use of suitable differential determinants which Hasse had used in the elliptic case. (See Sect. 8.3.1.) In any case, although Deuring announces that he will return to this question in the future he did not do so. (Compare the Remark on page 115.)

But Cartan was not able to check this since the latest Crelle volumes were missing in Strasbourg. Weil replied on 5 April 1940 that he is almost certain that Deuring has proved the said result for $h_n = n^{2g}$ with $n \not\equiv 0 \bmod p$ but how to be sure? In the meantime Weil will accept this fact, or prove it himself. He continues:

"My work makes striking progress. Almost all of the transcendental theory of algebraic functions can be transferred to finite fields of constants: period matrices, bilinear relations and the theorem of Hurwitz. The algebraic theory of Severi goes quite trivially. This has not been taken into account by Deuring who, like all these people who do not know their classics, has reformulated it in his language, with the tiresome notions of ideals, divisors, class field theory, residues, etc...."

Some days later, on 8 April 1940, Weil wrote again, saying that the absence of detailed information about the work of Hasse and Deuring is quite embarrassing to him; he needs not only the said result (namely $h_n = n^{2g}$ for $n \not\equiv 0 \bmod p$) but also some indication of their methods. He continues:

"Provisionally, while writing a note for the Comptes Rendus which I will send to your father I have decided to accept this fact for good without comment; but for my further work it would be useful for me to have some more precise information about the proof..."

Weil seems to be confident that the result is correct (what it is).

Henri Cartan's father was Elie Cartan who as a member of the academy was able to put the note into publication.

But then in his letter Weil mentions a more serious gap in his work, concerning what he calls his "fundamental lemma". He writes that he will complete his note even without having a proof of that lemma:

"I am quite clear about it so that I will take the risk ... Hasse does not have a chance any more,[2] *for I have solved (conditionally with respect to my lemma) all the main problems of the theory ..."*

Explicitly, Weil mentions first the problem of RHp (which Hasse had proved in the case of genus one), and secondly the problem that Artin's L-series for characters of Galois extensions are polynomials. He finds the second point more interesting since, he wrote, it opens the way for the study of non-abelian extensions in analogy to the situation of number fields. But the first point, he says, is more sensational in view of the tenacity with which Hasse and his people have tried to prove it.

Certainly, what Weil writes in this letter means a great success and he could rightly be proud of it. In the published version the note is registered as presented at the Academy session of 15 April 1940—about one week after this letter to Henri Cartan had been dispatched.

However, some time later his spirit seems to have calmed down. For on 2 May 1940 he writes:

"Since some time already I have left the algebraic functions which have continued to annoy me heavily. My main lemma is up in the air. In other situations I would have had scruples to publish it in this form, but in this moment this is not much of importance to me ... It may even turn out that what I am missing can be found in Deuring's work. When I am able to clear up this question I will return to it."

It is quite understandable that just in this moment the question about the lemma seemed not to be of primary importance to Weil in his personal situation. The court trial which would decide about his fate was due the next day.

Weil finally agreed to join the army under certain conditions. Some days later he could leave the prison. In a letter of 16 May 1940 to Cartan he wrote: *"Me voici soldat"*. In the turmoil created by the occupation of France by German troops which started on 10 May 1940 he succeeded to escape from France and finally enter the United States. He himself has told the details of his story in his autobiographic book [Wei91].

But the manuscript for Weil's CR-note had already been expedited. It appeared in print in the same year [Wei40]. Now, what precisely is the content of this note?

---

[2]*"Hasse n'a plus qu'a se prendre".*

12.2 The First Note 1940

The first sentence of Weil's note is rather long and reads:

"*I will briefly report in this note on the solution of the main problems of the theory of algebraic functions over finite base fields; it is known that these have been the topic of various papers, in particular during the last years by Hasse and his students; as they have anticipated, the key to these problems is the theory of correspondences; but the algebraic theory of correspondences which is due to Severi is not sufficient, and one has to add to these functions the transcendental theory of Hurwitz.*"

Then he reports about some details:

Given a function field $F_q|K_q$ with base field finite of cardinality $q = p^r$, consider its base field extension $F|K$ where K denotes the algebraic closure of K_q. Consider the subgroup $J^{(p)}$ of the elements of the Jacobian J of $F|K$ with order prime to p. Weil starts with the description of the structure of $J^{(p)}$, as a consequence of the formula $h_n = n^{2g}$ for $n \not\equiv 0 \mod p$. As he had written to H. Cartan he did not know whether this formula was proved already by Hasse and/or Deuring in characteristic p but nevertheless he uses it here (see the foregoing section). Accordingly the structure of $J^{(p)}$ is as follows:

Let $\widehat{\mathbb{Q}}^{(p)}$ denote the completion of the rational number field $\mathbb{Q}$ with respect to all ℓ-adic valuations except the p-adic one. $\widehat{\mathbb{Q}}^{(p)}$ contains the completion $\widehat{\mathbb{Z}}^{(p)}$. Then $J^{(p)}$ can be regarded as consisting of the vectors of length $2g$ with entries from the factor group $\widehat{\mathbb{Q}}^{(p)}/\widehat{\mathbb{Z}}^{(p)}$. That is, there is an isomorphism

$$J^{(p)} \approx \underbrace{\widehat{\mathbb{Q}}^{(p)}/\widehat{\mathbb{Z}}^{(p)} \times \cdots \times \widehat{\mathbb{Q}}^{(p)}/\widehat{\mathbb{Z}}^{(p)}}_{2g}$$

Every correspondence μ of the Jacobian J acts on this group and hence can be represented as a $2g \times 2g$ matrix $M(\mu)$ with entries in $\widehat{\mathbb{Z}}^{(p)}$.

Let μ' be the image of μ under the Rosati anti-automorphism. (The algebraic theory of the Rosati anti-automorphism had been developed by Deuring in his second paper on correspondences which appeared in the year [Deu40]. But probably Weil did not yet know about this paper.) Weil claims that the trace of the matrix $M(\mu\mu')$ is a rational integer and

$$\text{Tr}\, M(\mu\mu') > 0 .$$

This is the main content of the "important lemma" in Weil's note.

We see that Weil in this note follows precisely what Hasse had written to Weil in his letter of 12 July 1936 (see page 148). There, Hasse had said that one should search for a positive definite quadratic form on the ring of correspondences, and he had proposed that one should look for an abstract analogue of the Hermitian form belonging to the period matrix. Here, Weil points a way how to construct this form in analogy to the classical case, namely $\mu \mapsto \mathrm{Tr}\, M(\mu\mu')$. On first view this seems a good solution to Hasse's problem. But soon there arise questions which in this note are not suffiently explained. For instance, how did Weil make sure that the trace of $M(\mu\mu')$ is in $\mathbb{Z}$? It appears that in Weil's opinion this would allow him to use these matrices in the same way as in the classical case of characteristic 0 where the period matrices do have coefficients in $\mathbb{Z}$.

In his Collected Papers [Wei79a] Weil tells us that mathematically at "Bonne Nouvelle" he had to rely on his own recollections, i.e., he had no access to mathematical literature—with perhaps one exception:

> "It may be that I had at my disposal the stimulating book by Lefschetz "La Géometrie et l'Analysis Situs"; in any case this book kept my company a little later when I found myself in military uniform in May 1940."

Weil refers to the book [Lef24]. In fact, as far as one can see from the brief announcements in Weil's note [Wei40] the style of his note reminds me of Lefschetz's style in this book. But it seems futile to search for further stimuli which may have guided Weil during the preparation of his note. For, Weil himself explains in his recollections [Wei79a] in detail what ideas and thoughts were in his mind while writing up his CR-note.

REMARK The arrangement of Weil's Collected Papers, with the author's comments to each of his papers, is exemplary. It is an ideal source for the historian of mathematics who is interested in the development of ideas, theories and the mutual influence between the main actors. In this respect such arrangement is of much more value than the usual "Collected Papers" which contain the text of the various papers only. But it does not make superfluous the study of the letters which were exchanged at the time. Not infrequently it happens that after many years things in the past look somewhat different from what they formerly had been.

In any case, the main result of the note was "up in the air" as Weil had written to Henri Cartan in his letter of 2 May 1940 (see page 206). In the same letter he said that he will return to this topic when he has cleared up the open questions. This he did one year later, when he was already in the USA. There he published a second note about the RHp. (See the next section.)

But in the meantime Weil seems to have freely told people that in fact the publication of his first note [Wei40] was premature and he did not yet have a proof of the "important lemma". Deuring had said so in his paper [Deu41b], and he referred to "personal communication by the author". (See page 188.) I do not know how Deuring (in Germany) in those war times had been able to contact Weil (at that time in Princeton, USA). But in the summer of 1941 the USA were not yet directly involved in the war and perhaps the postal service still functioned. In a letter from Hasse to Julia dated 7 September 1941 Hasse informed Julia about the present state of Weil's "important lemma", i.e., that its validity was still unknown. (Julia had earlier sent a copy of Weil's CR-note to Hasse.) Hasse mentions Siegel and Chevalley, both in Princeton at that time, who could not verify the validity of the lemma and, when they had asked Weil he had admitted that he could not prove

it. I do not know how Hasse at that time had obtained this information. In a letter
to Dieudonné dated 13 March 1942 he wrote that he had heard this "indirectly"
from Siegel. Perhaps Hasse had in mind what Siegel earlier had written to him from
Princeton, in a letter dated 27 June 1940. Siegel had written:

> *"I have tried to understand the proof of the "Riemann hypothesis" for the zeta
> function of an algebraic function field with arbitrary genus with finite base
> field—the proof which André Weil has sketched in the "Comptes Rendus" of
> 22 April. But right at the beginning I encountered a difficulty which I could
> not overcome. I had asked Chevalley about it but he couldn't do it either. But
> you will probably be able to do it. . . "*

However, from the date of this letter it is evident that at that time Weil had not
yet arrived in the USA and therefore he could not have been consulted by Siegel.
It appears that Hasse later had found another source for his information. Perhaps
Deuring? In any case it seems to have been generally known to the people involved,
including the referees, that the "important lemma" was still up in the air. The
referee in the "*Jahrbuch*" for Weil's first note was van der Waerden. The referee
in the "*Zentralblatt*" was H. L. Schmid. In the "*Mathematical Reviews*" it was
O.F.G. Schilling.

Quite generally, it is tacitly assumed that a note in the "Comptes Rendus" (or
in the notices of any scientific academy in the world) represents a preliminary
announcement of results which, at least in the opinion of the author, have already
been obtained—if not explicitly stated otherwise. Weil in the first sentence of the
note gives the impression that indeed he had already obtained "*the solution of the
main problems of the theory of algebraic functions over finite base fields*". Hence
it is understandable that Hasse and others, after hearing that the main lemma was
still unproved, got the impression that this note did not follow the usual standards
of scientific publishing. In the letter to Julia cited above, Hasse voiced his suspicion
that Weil may have written this note in order to secure priority by unfair methods.
In a similar vein he expressed himself later in a letter to Dieudonné dated 13 March
1942. (Dieudonné had informed him in a letter of 27 February 1942 that in the
meantime Weil had found a complete proof of the RHp. Obviously he referred
to Weil's second note [Wei41]; see next section.) Hasse did not know, and in any
case did not take into consideration the quite extreme situation which Weil was
confronted with when he decided to have this note published. Weil himself in
[Wei79a] gave this as an explanation for his premature publication.

12.3 The Second Note 1941

Some time in the summer of 1941 Hasse received mail from the USA. The sender
was André Weil. He sent a reprint of a second note, this time published in the
Proceedings of the National Academy of Sciences, announcing the proof of the

RHp [Wei41]. This was accompanied by a short letter (in English language) with the following text:

> *"Dear Hasse, In the midst of the vastly more important affairs in which I hear you are at present engaged, you may still be able to spare a few minutes for the perusal of the solution of a problem you used to be interested in. With best greetings from the U. S. A. Yours sincerely A Weil."*

REMARK I have not found this letter among the Hasse papers. Hasse had informed Julia about Weil's second note and at that occasion cited this text, in his letter to Julia dated 7 September 1941. I have copied this text from Hasse's letter to Julia. The somewhat sarcastic tone of Weil's text sounds different from that of the former letters of Weil to Hasse. This may perhaps be explained with the different political attitudes of these two people in that time. It was World War II. Weil was strictly against military service and a fugitive from the German terror; Hasse served in a military research group following his patriotic sentiment.

Weil had arrived as a refugee in the United States in January 1941. There he had close contact to his friend Chevalley and other people in Princeton. It seems he worked hard in order to prove the RHp and thus to vindicate his suspicious note of last year. He succeeded quickly and published the above mentioned second note which carries the date 11 June 1941. There he again mentions

> *"... the two outstanding problems, viz. the proof of the Riemann hypothesis for such fields and the proof that Artin's nonabelian L-functions on such fields are polynomials."*

Here, "such fields" are algebraic function fields of one variable with finite base fields.

In the first paragraph of the note Weil refers to his earlier CR-note of last year where, he says, he had

> *"sketched the outline of a new theory of algebraic functions over a finite field of constants"*

which might be useful to prove the two problems above. That new theory may be described as "transcendental", says Weil, since it uses close analogues to the classical theory of abelian integrals of the first kind and Jacobi's Inversion Theorem. But:

> *"...I have now found that my proof of these results is independent of that "transcendental" theory and depends only upon the algebraic theory of correspondences on algebraic curves, as due to Severi."*

In other words: Now he followed strictly the direction which Deuring had set with his theory of correspondences and which Hasse had informed him about in the year 1936 (see Sects. 9.3 and 9.4). To be sure: the mathematical languages used by Weil and Deuring are different: Weil is using the language of algebraic geometry and Deuring that of algebraic function fields. But both are belonging to algebra: Deuring is the translation from the geometric language into the language of function fields, and Weil keeps the geometric language following Severi, but with the following comment:

> *"It should be observed, however, that Severi's treatment, although undoubt-edly containing all the essential elements for the solution of the problems it purports to solve, is meant to cover only the classical case where the field of constants is that of complex numbers, and doubts may be raised as to its applicability to more general cases, especially to characteristic $p \neq 0$. A rewriting of the whole theory, covering such cases, is therefore a necessary preliminary to the applications we have in view."*

Here we see already Weil's plan to rewrite the whole of classical algebraic geometry for arbitrary base fields—a huge effort which he finally mastered in his book [Wei46].

But apart from the mathematical language, Weil goes one essential step further than Deuring. Whereas Deuring stopped after constructing algebraically the ring of endomorphisms of the Jacobian, Weil followed Severi further and found the positive definite quadratic form on it which had been envisaged by Hasse. When Weil said that Severi's treatment "undoubtedly contains all the essential elements..." this is only a mild circumlocution of saying that Severi's arguments are to be cleared and rewritten such that they meet the standards of a modern mathematical exposition. And this is precisely what Weil does. In particular this concerns the proof that the quadratic form in question is positive definite—which is the main point for the RHp.

In my opinion, this is the essential achievement of Weil with respect to our story of the RHp. He was the one who did what nobody had been able to do, namely he dived into the ocean of algebraic geometry and brought to the surface the sources which were needed for the proof of RHp, in particular a consistent theory of intersection number. As an extra bonus he provided a new foundation of the whole of algebraic geometry for arbitrary characteristic, which required a huge effort and started a new era of "arithmetic geometry". This would not have been absolutely necessary just for the proof of the RHp. For, that proof is "elementary", as Weil says in his letter to Artin [Wei80], since it uses only the geometry on the direct product $\Gamma \times \Gamma$ of a curve with itself.

Now, how does Severi's proof look like after Weil's rewriting? This I have pointed out already in Sect. 10.5. You have just to read our "virtual proof" in the language of algebraic geometry, in the way as I indicated there. The important part is the proof of positive definiteness (Sect. 10.2). I admit that the proof there is not exactly the rewriting of Weil's proof but, I believe, shows some simplifications. But the main arguments there are essentially modeled after Weil's. In Weil's proof appears what has been called the "generic complementary correspondence" which I have avoided in the virtual proof and replaced by a simple discriminant estimate. See Sect. 10.2.2.

REMARK As it turned out, this quadratic form and its positivity had already been found earlier in classical algebraic geometry, by Castelnuovo. This was later discovered and pointed out by A. Weil in [Wei56]. Weil mentions that Castelnuovo had defined algebraically for any correspondence A what he called "equivalence defect" $\delta(A) \geq 0$ which is constructed precisely as our $\sigma(A, A)$ and vanishes if and only if $A \approx 0$. Apparently this was not known by the participants of Hasse's workshop, nor to Weil at that time. Today the statement $\sigma(A, A) > 0$ is called "Inequality of Castelnuovo-Severi".

Weil's other proof, which he called "transcendental" and which "was up in the air" at the time of his first note, did still remain up there when he published his second note. But he promised at the end of this note:

"A detailed account of this theory, including ... the "transcendental" theory as outlined in my previous note, is prepared for publication."

In consequence Weil published the books [Wei46, Wei48a, Wei48b].

Prior to these publications, after Weil had finally convinced himself about his proofs and before going to prepare the final manuscripts, he wrote a letter to Artin explaining all details. The letter is dated 10 July 1942, it begins with:

"I have now reached a state in my work on correspondences where it will be helpful if I make a general survey of the theory, for you and a few such people. This seems all the more desirable, as I now find that the final writing up is going to involve a recasting of the intersection theory in algebraic geometry (since I am not altogether pleased with v. d. Waerden's treatment of this subject), this means that things may not get ready for quite a while."

The full text of the letter is contained in the "Collected papers" of Weil, as well as his comments. I have the impression that the letter was written not only as information for Artin and other "such people", but also as an opportunity for Weil to put his ideas into line and to recheck the whole setup.

The fact that the letter was written to Artin shows that Artin still was considered as the ultimate authority for the RHp, although he had not published anything about this question after his thesis.

Summary

After the outbreak of the second world war (1939) Hasse was drafted to the Navy and did not have much time to think about the RHp. Weil found himself imprisoned in Rouen as a conscientious objector to military service (1940). While waiting for the trial he used his time to think about mathematics, in particular about a proof of the RHp. This is documented in the letters which he exchanged with Henri Cartan and which are preserved. After a while he had an idea how to algebraize the transcendental theory of correspondences, for which he cited Hurwitz, and to find the positive definite form which Hasse had envisaged in a letter to him. However, since in his prison he had no access to mathematical literature he was not able to check the details. But although the proof was still incomplete he decided, due to his very exceptional situation, to publish a note in the "Comptes Rendus" explaining his ideas.

During the turmoil created by the German invasion of France Weil managed to escape to the USA. There he continued his work on the RHp and in 1941 he succeeded with a proof of the RHp. However this new proof went differently from the one which he had sketched in his CR-note. It used the algebraic theory of correspondences as developed by Severi. He was able to write a new foundation of algebraic geometry, containing in particular the intersection theory of algebraic varieties. His results were finally published in three books 1946/48. The essential theorem leading to the RHp is what today is called "Inequality of Castelnuovo-Severi".

In the year 1942, hence prior to the above mentioned publications, Weil wrote a letter to Artin explaining all details.

Chapter 13
Appendix

With the appearance of Weil's above mentioned three books, the RHp[1] was settled and our story comes to an end. But the mathematical development inspired by this or that item of our story persists and is still present. From the numerous literature in this direction I will mention here three papers only:

The first is the paper of Mattuck and Tate published in the year 1958 in the *Hamburger Abhandlungen* [MT58]. There the authors show that the RHp can be derived from the Riemann-Roch Theorem for surfaces, applied to the surface $\Gamma \times \Gamma$, the direct product of a curve Γ with itself. This is an interesting aspect. It has inspired Grothendieck to investigate also for higher-dimensional varieties the quadratic form on the Neron-Severi group, defined by the intersection multiplicities of divisors [Gro58].

Secondly, I have to mention Weil's generalization of the RHp to varieties of higher dimension, culminating in the so-called "Weil conjectures". We have seen the beginning of these ideas in the letter of Weil to Hasse dated 24. February 1939 (see page 194). Weil's conjectures were finally solved by Déligne. This is again a fascinating story but it would exceed the scope of this book. I refer to the presentation by Freitag and Kiehl in the book [FK88] which includes a historic sketch by Dieudonné.[2]

---

[1] And much more.

[2] Added in proof: I am indebted to Franz Lemmermeyer for pointing out to me the recent papers [Mil16, OS16].

© Springer Nature Switzerland AG 2018

P. Roquette, *The Riemann Hypothesis in Characteristic* p
in Historical Perspective, Lecture Notes in Mathematics 2222,
https://doi.org/10.1007/978-3-319-99067-5_13

13.1   Bombieri

Thirdly, there has appeared a new and quite short proof of the RHp by Bombieri [Bom74]. This came as a surprise to the mathematical community since the underlying idea is different from the original ideas of Hasse, Deuring and Weil. Those people had seen the analogy of the RHp problem to the classic theories of complex multiplication, of analysis and algebraic geometry. They had successfully remodelled the relevant parts of those classic theories such as to cover also the case of characteristic $p > 0$. In particular they could now handle the Frobenius operator which, due to his inseparability properties, does not have an analogue in the characteristic 0 case and in this sense may look as a somewhat strange exemption although it plays the main role in RHp. In constrast, Bombieri does not care about the historic analogies to the characteristic 0 case. Following a lead from Stepanov [Ste69] he uses explicitly the strength of the inseparability properties of the Frobenius operator in order to reach the RHp directly.

Originally I had planned this last chapter as a brief epilogue only. But now I cannot resist to include here at least a sketch of the main idea of this short and highly original piece of work. I am closely following Bombieri's presentation but use my own notation. Let us review the situation:

$F|K$ function field with finite base field.
  q number of elements of the base field K.
  g the genus of $F|K$.
  N the number of primes of degree 1 of $F|K_q$.

It is not necessary here to work in the base field extension with the algebraic closure of K as new base field, as was done in earlier sections. The main result of Bombieri's new proof is the

Theorem *Suppose that $q = p^{2\mu}$ is a square, and that q is sufficiently large. Then*

$$N \leq q + (2g + 1)\sqrt{q} \,. \tag{13.1}$$

When I say that q should be "sufficiently large" then this means that q should be larger than a constant depending on the genus g only. The following proof will produce such a constant. The assumptions on q can be achieved after suitable base field extension of the given function field.

On first view the theorem looks like the verification of Artin's criterion (see page 52 in Sect. 4.5). But it is only part of it. Artin's criterion requires the estimate of $|N - q - 1|$ if q is sufficiently large. Therefore, in addition to (13.1) it will be necessary to obtain a suitable lower estimate too. But once one has the idea leading to the above theorem, the required lower estimate can be obtained by applying the same idea to other situations, which appear when the underlying meromorphism is replaced by its inverse in the (additive) group of endomorphisms of the Jacobian of $F|K$. Therefore I will concentrate here on the above theorem. Bombieri's full proof has been included into the well known book "Field Arithmetic" by Jarden and Fried [FJ08].

In the former proofs the idea of obtaining an estimate of N was based on the construction of an element $0 \neq z \in F$ which has all primes of degree 1 as zeros. Then N can be estimated by the degree of the zero divisor of z. This equals the degree of the pole divisor of z. Thus one is left with estimating the degree of this pole divisor. Compare, e.g., our construction of the determinant $d_\mu(u)$ on page 171 f. There we have estimated the degree of a discriminant, which is the square of a different. Note that N can be interpreted as the degree of the different $\mathfrak{D}(1, \pi)$; see page 174. But that estimate above is quite rough since the pole divisor of the determinant in question is quite large. That method had given the desired result only if embedded as part of the general theory of correspondences.

The new idea of Stepanov-Bombieri is to construct directly a non-zero element z in F which again has the primes of degree 1 as zeros *but this time of high multiplicity*. It will turn out that the multiplicities of the zeros of z are to be $\geq p^\mu = \sqrt{q}$; thus $p^\mu N$ will be estimated by the degree of the pole divisor of z. This will give a better estimate for N and leads to the desired result.

Actually, in Bombieri's construction the pole divisor of z will be a power of a prime P_0 of degree 1 which is chosen in advance. Thus Bombieri's construction will give an estimate of the number $N - 1$ of the remaining primes of degree 1.

The details of this construction are as follows:

P_0 is a prime of degree 1 of $F|K$.

$\mathcal{L}_\ell = \mathcal{L}(P_0^\ell)$ the module of elements in F having P_0 as their only pole, of order $\leq \ell$.

$\mathcal{L}_\ell^{p^\mu}$ the module consisting of the p^μ-th powers of elements of $\mathcal{L}_\ell$.

The desired element z is to be contained in the product module

$$\mathcal{L}_\ell^{p^\mu} \cdot \mathcal{L}_m^{p^{2\mu}}$$

where ℓ, m are certain parameters which will have to be carefully selected (see below). The pole divisor of any $z \in \mathcal{L}_\ell^{p^\mu} \cdot \mathcal{L}_m^{p^{2\mu}}$ is a power of P_0, of degree $\leq \ell p^\mu + m p^{2\mu}$. Every $z \in \mathcal{L}_\ell^{p^\mu} \cdot \mathcal{L}_m^{p^{2\mu}}$ is a p^μ-th power and therefore every zero of z has multiplicity $\geq p^\mu$. Thus we only have to make sure that there exists some $z \in \mathcal{L}_\ell^{p^\mu} \cdot \mathcal{L}_m^{p^{2\mu}}$ which admits all $N - 1$ primes $\neq P_0$ of degree 1 as zeros. If this is achieved then, as explained above, we obtain the estimate

$$p^\mu(N - 1) \leq \ell p^\mu + m p^{2\mu}$$

$$N - 1 \leq \ell + m p^\mu.$$

Choosing the parameters

$$\ell = p^\mu - 1, \qquad m = p^\mu + 2g \tag{13.2}$$

we obtain the announced estimate (13.1) since $p^{2\mu} = q$.

This choice of parameters may look somewhat artificial at this point but the following computations will show how they arise when one is looking for the existence of $z \in \mathcal{L}_\ell^{p^\mu} \cdot \mathcal{L}_m^{p^{2\mu}}$ which has all primes of degree 1 (except P_0) as zeros.

If $u_1, u_2, u_3 \ldots$ is a K-basis of $\mathcal{L}_\ell$ then every $z \in \mathcal{L}_\ell^{p^\mu} \cdot \mathcal{L}_m^{p^{2\mu}}$ can be represented in the form:

$$z = \sum_i u_i^{p^\mu} x_i^{p^{2\mu}} \qquad \text{with } x_i \in \mathcal{L}_m. \tag{13.3}$$

Since $\ell < p^\mu$ by (13.2) this representation is unique. To see this choose the basis of $\mathcal{L}_\ell$ adapted to the series of submodules

$$K = \mathcal{L}_0 \subset \mathcal{L}_1 \subset \cdots \subset \mathcal{L}_{\ell-1} \subset \mathcal{L}_\ell.$$

Each $\mathcal{L}_i$ is of dimension ≤ 1 over the preceding module; if that dimension is 1 then choose $u_i \in \mathcal{L}_i \setminus \mathcal{L}_{i-1}$. This gives a K-basis of $\mathcal{L}_\ell$, and $v_{P_0}(u_i) = -i$. Since $i \le \ell < p^\mu$ the $v_{P_0}(u_i)$ are mutually incongruent modulo p^μ. Hence the u_i are linearly independent over F^{p^μ}, and therefore the $u_i^{p^\mu}$ are linearly independent over $F^{p^{2\mu}}$. Thus indeed, the representation (13.3) is unique. In other words: $\mathcal{L}_\ell^{p^\mu} \cdot \mathcal{L}_m^{p^{2\mu}}$ is isomorphic to the tensor product $\mathcal{L}_\ell^{p^\mu} \otimes \mathcal{L}_m^{p^{2\mu}}$ over K.

In order to make sure that the element z in (13.3) has all primes of degree 1 as zeros (except P_0) one tries to choose z such that

$$\widetilde{z} := \sum_i u_i^{p^\mu} \cdot x_i = 0.$$

In fact, if $P \neq P_0$ is of degree 1 then $x_i P \in K$ for every i and hence

$$x_i^{p^{2\mu}} P = (x_i P)^{p^{2\mu}} = (x_i P)^q = x_i P.$$

Observe that the map $x \mapsto x^{p^{2\mu}}$ is the Frobenius meromorphism π of $F|K$ since $p^{2\mu} = q$. (Compare (7.12) on page 94). It follows

$$zP = \sum_i u_i^{p^\mu} P \cdot x_i^{p^{2\mu}} P = \sum_i u_i^{p^\mu} P \cdot x_i P = \widetilde{z}P = 0.$$

The map $z \mapsto \widetilde{z}$ is a K-homomorphism

$$\mathcal{L}_\ell^{p^\mu} \otimes \mathcal{L}_m^{p^{2\mu}} \longrightarrow \mathcal{L}_\ell^{p^\mu} \cdot \mathcal{L}_m \subset \mathcal{L}_{\ell\, p^\mu + m}. \tag{13.4}$$

Thus $z \neq 0$ has to be in the kernel of this map. But:

What if the kernel of the map $z \mapsto \widetilde{z}$ vanishes?

Well, if this would be the case then the proof would break down. So we have to look for an argument that indeed the kernel does not vanish for the above choices of the parameters ℓ and m in (13.2)—provided q is sufficiently large. At this point the Riemann-Roch theorem comes into the play.

The Riemann-Roch Theorem allows to estimate the K-dimensions of the respective modules in question:

$$\dim \mathcal{L}_\ell^{p^\mu} = \dim \mathcal{L}_\ell \geq \ell - g + 1$$

$$\dim \mathcal{L}_m^{p^{2\mu}} = \dim \mathcal{L}_m \geq m - g + 1$$

$$\dim \mathcal{L}_{\ell\, p^\mu + m} = \ell\, p^\mu + m - g + 1 \quad \text{(if } \ell p^\mu + m > 2g - 2 \text{)}.$$

Thus the kernel of $z \mapsto \tilde{z}$ does not vanish if $\ell p^\mu + m > 2g - 2$ and

$$(\ell - g + 1)(m - g + 1) > \ell\, p^\mu + m - g + 1.$$

Substituting for ℓ, m the terms in (13.2) shows $\ell p^\mu + m = p^{2\mu} + 2g$, and gives the condition

$$p^{2\mu} + p^\mu - g(g+1) > p^{2\mu} + g + 1,$$

$$p^\mu > (g+1)^2$$

$$q = p^{2\mu} > (g+1)^4.$$

This is meant in the theorem when I said that q should be "suffiently large".

I hope I have been able to convey to you the originality and beauty of the idea for this new proof. But at the same time it is seen that the realization of the idea requires a subtle choice of the parameters involved in the construction, see (13.2). It is not at all clear "why" just these or similar choices lead to the envisaged result.

What would Hasse have said if he had seen this proof? Bombieri's proof appeared in 1976; Hasse died in the year 1979. I do not know whether Hasse in his last years had seen or at least been told about Bombieri's paper. In any case, if I am allowed to speculate then I would say that he would have fully acknowledged the achievement of the author and its ingenuity. But on the other hand I recall Hasse's discussion with Davenport about proofs as manifestations of mathematical structures. (See page 64). Perhaps he would have added that one should search for the "true" reason, embodied in the structure of function fields and in particular the Riemann-Roch theorem. How can one explain that the Bombieri-Stepanov idea has been so successful in exhibiting just $p^\mu = \sqrt{q}$ as the error term? What is its connection to the idea of Hasse, Deuring and Weil who constructed the quadratic form? In my opinion such investigation would be a good and interesting problem. Note that the Riemann-Roch theorem, which is essentially the basis of Bombieri's proof, is equivalent to the functional equation of the zeta function $\zeta_F(s)$. This had been shown by F. K. Schmidt and Witt. (See Sect. 4.4). Is it possible to derive (13.1) directly from the functional equation of $\zeta_F(s)$?

Correction to: The Riemann Hypothesis in Characteristic p in Historical Perspective

Correction to:
P. Roquette, *The Riemann Hypothesis in Characteristic* p *in Historical Perspective*, **Lecture Notes in Mathematics 2222,**
https://doi.org/10.1007/978-3-319-99067-5

The book was inadvertently published without updating the following photo source information:

Photo of Emil Artin, outside his apartment in Hamburg, approx. 1935 on page 13
© Estate of Natascha Artin-Brunswick

Photo of Friedrich Karl Schmidt on page 39
Author: Konrad Jacobs, Source © Archives of the Mathematisches Forschungsinstitut Oberwolfach

Photo of Helmut Hasse, Hamburg, approx. 1934 on page 55
Peter Roquette's personal photo archive

Photo of Harold Davenport, Cambridge approx. 1932 on page 59
© James H. Davenport

Photo of Max Deuring, 1973 on page 136
Author Konrad Jacobs, Source © Archives of the Mathematisches Forschungsinstitut Oberwolfach

© Springer Nature Switzerland AG 2018
P. Roquette, *The Riemann Hypothesis in Characteristic* p
in Historical Perspective, Lecture Notes in Mathematics 2222,
https://doi.org/10.1007/978-3-319-99067-5_14

Photo of Andre Weil on page 203
© Sylvie Weil personal archives

The source lines have been added to the book FM pages.

The updated online version of the book can be found at
https://doi.org/10.1007/978-3-319-99067-5

References

[AH25] E. Artin, H. Hasse, Über den zweiten Ergänzungssatz zum Reziprozitätsgesetz der l–10 Potenzreste im Körper k_ζ der l-ten Einheitswurzeln und in Oberkörpern von k_ζ. J. Reine Angew. Math. **154**, 143–148 (1925)

[Art00] E. Artin, Quadratische Körper über Polynombereichen Galoisscher Felder und ihre Zetafunktionen. Edited by Peter Ullrich. Abh. Math. Semin. Univ. Hamburg **70**, 3–30 (2000)

[Art21] E. Artin, Quadratische Körper im Gebiete der höheren Kongruenzen. Jahrbuch der Philosophischen Fakultät zu Leipzig **1921**, 157–165 (1921)

[Art24] E. Artin, Quadratische Körper im Gebiete der höheren Kongruenzen I, II. Math. Zeitschr. **19**, 153–206, 207–246 (1924)

[Art65] E. Artin, *The Collected Papers of Emil Artin*, ed. by S. Lang, J.T. Tate (Addison-Wesley, Reading, MA, 1965). XVI+560 pp.

[AS27a] E. Artin, O. Schreier, Algebraische Konstruktion reeller Körper. Abh. Math. Semin. Univ. Hamburg **5**, 85–99 (1927)

[AS27b] E. Artin, O. Schreier, Eine Kennzeichnung der reell abgeschlossenen Körper. Abh. Math. Semin. Univ. Hamburg **5**, 225–231 (1927)

[Aud11] M. Audin, Correspondance entre Henri Cartan et André Weil (1928–1991). Paris: Société Mathématique de France, 2011. xiii, 720 pp.

[AW45] E. Artin, G. Whaples, Axiomatic characterization of fields by the product formula for valuations. Bull. Am. Math. Soc. **51**, 469–492 (1945)

[Beh35] H. Behrbohm, Über die Algebraizität der Meromorphismen eines elliptischen Funktionenkörpers. Nachr. Ges. Wiss. Göttingen (2) **1**, 131–134 (1935)

[BM04] J. Brillhart, P. Morton, Class number of quadratic fields, Hasse invariants of elliptic curves, and the supersingular polynomial. J. Number Theory **106**, 79–111 (2004)

[Bom74] E. Bombieri, Counting points on curves over finite fields (d'apres S.A. Stepanov), *Sem. Bourbaki 1972/1973*. Expose No. 430, Lecture Notes in Mathematics, vol. 383 (Springer, Berlin, 1974), pp. 234–241

[BR36] H. Behrbohm, L. Redei, Der Euklidische Algorithmus in quadratischen Körpern. J. Reine Angew. Math. **174**, 192–205 (1936)

[Cea96] S.S. Chern et al., Wei-Liang Chow 1911–1995. Notices Am. Math. Soc. **43**(10), 1117–1124 (1996)

[Che36] C. Chevalley, Généralisation de la théorie du corps de classes pour les extensions infinies. J. Math. pur. Appl. (9) **15**, 359–371 (1936)

© Springer Nature Switzerland AG 2018

P. Roquette, *The Riemann Hypothesis in Characteristic* p
in Historical Perspective, Lecture Notes in Mathematics 2222,
https://doi.org/10.1007/978-3-319-99067-5

[Che40] C. Chevalley, La théorie du corps de classes. Ann. Math. (2) **41**, 394–418 (1940)

[Che51] C. Chevalley, *Introduction to the Theory of Algebraic Functions of One Variable.* Mathematical Surveys, vol. VI (American Mathematical Society, New York, 1951) XI, 188 pp.

[CN35] C. Chevalley, H. Nehrkorn, Sur les démonstrations arithmétiques dans la théorie du corps de classes. Math. Ann. **111**, 364–371 (1935)

[CW34] C. Chevalley, A. Weil, Über das Verhalten der Integrale 1. Gattung bei Automorphismen des Funktionenkörpers. Abh. Math. Semin. Univ. Hamburg **10**, 358–361 (1934)

[Dav31] H. Davenport, On the distribution of quadratic residues (mod p). J. Lond. Math. Soc. **6**, 49–54 (1931)

[Dav33] H. Davenport, On the distribution of quadratic residues (mod p). (Second paper.). J. Lond. Math. Soc. **8**, 46–52 (1933)

[Dav36] H. Davenport, The meromorphisms of an elliptic function field. Proc. Camb. Philos. Soc. **32**, 212–215 (1936)

[Dav77] H. Davenport, *The Collected Works I–IV*, ed. by B.J.Birch, H. Halberstam, C.A. Rogers (Academic, New York, 1977)

[Ded57] R. Dedekind, Abriss einer Teorie der höheren Kongruenzen in Bezug auf einen reelen Primzahl-Modulus. J. Reine Angew. Math. **54**, 1–26 (1857). Nachdruck in Gesammelte mathematische Werke, Erster Band, V, pp. 40–67

[Deu35] M. Deuring, Algebren. Erg. d. Math. u. ihrer Grenzgebiete. Julius (Springer, Berlin, 1935), 143 pp.

[Deu36] M. Deuring, Automorphismen und Divisorklassen der Ordnung ℓ in algebraischen Funktionenkörpern. Math. Ann. **113**, 208–215 (1936)

[Deu37] M. Deuring, Arithmetische Theorie der Korrespondenzen algebraischer Funktionenkörper. I. J. Reine Angew. Math. **177**, 161–191 (1937)

[Deu40] M. Deuring, Arithmetische Theorie der Korrespondenzen algebraischer Funktionenkörper. II. J. Reine Angew. Math. **183**, 25–36 (1940)

[Deu41a] M. Deuring, Die Typen der Multiplikatorenringe elliptischer Funktionenkörper. Abh. Math. Semin. Univ. Hamburg **14**, 197–272 (1941)

[Deu41b] M. Deuring, La teoria aritmetica delle funzioni algebriche di una variabile. Rend. Mat. Appl. V. Ser. **2**, 361–412 (1941)

[Deu42] M. Deuring, Reduktion algebraischer Funktionenkörper nach Primdivisoren des Konstantenkörpers. Math. Z. **47**, 643–654 (1942)

[Deu49] M. Deuring, Algebraische Begründung der komplexen Multiplikation. Abh. Math. Semin. Univ. Hamburg **16**(1/2), 32–47 (1949)

[Deu55] M. Deuring, Die Zetafunktion einer algebraischen Kurve vom Geschlechte eins. II. Nachr. Akad. Wiss. Göttingen, Math.-Phys. Kl. **1955**, 13–42 (1955)

[DH34] H. Davenport, H. Hasse, Die Nullstellen der Kongruenzzetafunktionen in gewissen zyklischen Fällen. J. Reine Angew. Math. **172**, 151–182 (1934)

[DS15] D. Dumbaugh, J. Schwermer, *Emil Artin and Beyond. Class Field Theory and L-Functions. With Contributions by James Cogdell and Robert Langlands* (European Mathematical Society (EMS), Zürich, 2015)

[DW82] R. Dedekind, H. Weber, Theorie der algebraischen Funktionen einer Veränderlichen. J. Reine Angew. Math. **92**, 181–290 (1882)

[Eic37] M. Eichler, Über die Idealklassenzahl total definiter Quaternionenalgebren. Math. Z. **43**, 102–109 (1937)

[FJ08] M.D. Fried, M. Jarden, *Field Arithmetic*, 3rd edn. (Springer, Berlin, 2008), xiv+792 pp. Revised edn. by Moshe Jarden (2008)

[FK88] E. Freitag, R. Kiehl, *Étale Cohomology and the Weil Conjecture. With a Historical Introduction by J. A. Dieudonné* (Springer, Berlin, 1988)

[FLR14] G. Frei, F. Lemmermeyer, P. Roquette, (eds.), *Emil Artin and Helmut Hasse. Their Correspondence 1923–1958* English version, revised and enlarged. Contributions in Mathematical and Computational Science, vol. 5 (Springer Basel, Cham, 2014) X + 484 pp.

[FR08] G. Frei, P. Roquette (eds.), *Emil Artin and Helmut Hasse. Their correspondence 1923–1934. With contributions of Franz Lemmermeyer and an introduction in English.* (Universitäts–Verlag, Göttingen, 2008), 497 pp.

[Fre04] G. Frei, On the history of the artin reciprocity law in abelian extensions of algebraic number fields: How Artin was led to his reciprocity law, in *The Legacy of Niels Henrik Abel. The Abel Bicentennial, Oslo 2002*, ed. by O.A. Laudal, R. Piene (Springer, Berlin, 2004)

[Gro58] A. Grothendieck, Sur une note de Mattuck-Tate. J. Reine Angew. Math. **200**, 208–215 (1958)

[Has02] H. Hasse, *Number Theory*. Transl. from the 3rd German edn. edited and with a preface by Horst Günter Zimmer. Reprint of the 1980 edn. (Springer, Berlin, 2002), 638 pp.

[Has24] H. Hasse, Darstellbarkeit von Zahlen durch quadratische Formen in einem beliebigen algebraischen Zahlkörper. J. Reine Angew. Math. **153**, 113–130 (1924)

[Has26a] H. Hasse, Bericht über neuere Untersuchungen und Probleme aus der Theorie der algebraischen Zahlkörper. I: Klassenkörpertheorie. Jahresber. Dtsch. Math.-Ver. **35**, 1–55 (1926)

[Has26b] H. Hasse, Neue Begründung der komplexen Multiplikation I: Einordnung in die allgemeine Klassenkörpertheorie. J. Reine Angew. Math. **157**, 115–139 (1926)

[Has27] H. Hasse, Bericht über neuere Untersuchungen und Probleme aus der Theorie der algebraischen Zahlkörper. Teil Ia: Beweise zu I. Jahresber. Dtsch. Math.-Ver. **36**, 233–311 (1927)

[Has30] H. Hasse, Bericht über neuere Untersuchungen und Probleme aus der Theorie der algebraischen Zahlkörper. II: Reziprozitätsgesetz. Jahresber. Dtsch. Math.-Ver., 6(Ergänzungsband), 1930. IV + 204 pp.

[Has31] H. Hasse, Neue Begründung der komplexen Multiplikation. II. Aufbau ohne Benutzung der allgemeinen Klassenkörpertheorie. J. Reine Angew. Math. **165**, 64–88 (1931)

[Has32] H. Hasse, Theory of cyclic algebras over an algebraic number field. Trans. Am. Math. Soc. **34**, 171–214 (1932)

[Has33a] H. Hasse, Beweis des Analogons der Riemannschen Vermutung für die Artinschen und F. K. Schmidtschen Kongruenzzetafunktionen in gewissen elliptischen Fällen. Vorläufige Mitteilung. Nachr. Ges. Wiss. Göttingen, Math.–Phys. Kl. I **1933**(42), 253–262 (1933)

[Has33b] H. Hasse, Vorlesungen über Klassenkörpertheorie. Preprint, Marburg. [Later published in book form by Physica Verlag Würzburg (1967)] (1933)

[Has34a] H. Hasse, Abstrakte Begründung der komplexen Multiplikation und Riemannsche Vermutung in Funktionenkörpern. Abh. Math. Semin. Univ. Hamburg **10**, 325–348 (1934)

[Has34b] H. Hasse, Existenz separabler zyklischer unverzweigter Erweiterungskörper vom Primzahlgrad p über elliptischen Funktionenkörpern der Charakteristik p. J. Reine Angew. Math. **172**, 77–85 (1934)

[Has34c] H. Hasse, Riemannsche Vermutung bei den F.K.Schmidtschen Kongruenzzetafunktionen. Jahresber. Dtsch. Math.-Ver. **44**, 2.Abt., 44 (1934)

[Has34d] H. Hasse, Theorie der Differentiale in algebraischen Funktionenkörpern mit vollkommenem Konstantenkörper. J. Reine Angew. Math. **172**, 55–64 (1934)

[Has34e] H. Hasse, Theorie der relativ–zyklischen algebraischen Funktionenkörper, insbesondere bei endlichem Konstantenkörper. J. Reine Angew. Math. **172**, 37–54 (1934)

[Has34f] H. Hasse, Über die Kongruenzzetafunktionen. Unter Benutzung von Mitteilungen von Prof. Dr. F. K. Schmidt und Prof. Dr. E. Artin. Sitzungsber. Preuß. Akad. Wiss., Phys.-Math. Kl. **1934**(17), 250–263 (1934)

[Has35] H. Hasse, Zur Theorie der abstrakten elliptischen Funktionenkörper. Nachr. Ges. Wiss. Göttingen I. N.F. **1**, 119–129 (1935)

[Has36a] H. Hasse, Theorie der höheren Differentiale in einem algebraischen Funktionenkörper mit vollkommenem Konstantenkörper der Charakteristik p. J. Reine Angew. Math. **175**, 50–54 (1936)

[Has36b] H. Hasse, Zur Theorie der abstrakten elliptischen Funktionenkörper. III. die Struktur des Meromorphismenrings. J. Reine Angew. Math. **175**, 193–207 (1936)

[Has36c] H. Hasse, Zur Theorie der abstrakten elliptischen Funktionenkörper. I,II,III. J. Reine Angew. Math. **175**, 55–62, 69–81, 193–207 (1936)

[Has37a] H. Hasse, Anwendungen der Theorie der algebraischen Funktionen in der Zahlentheorie. Abh. Ges. Wiss. Göttingen. Math.-Phys. Klasse, III. Folge(Heft 18), 51–55 (1937)

[Has37b] H. Hasse, Die Gruppe der p^n-primären Zahlen für einen Primteiler p von p. J. Reine Angew. Math. **176**, 174–183 (1937)

[Has37c] H. Hasse, Noch eine Begründung der Theorie der höheren Differentialquotienten in einem algebraischen Funktionenkörper einer Unbestimmten. Nach einer brieflichen Mitteilung von F. K. Schmidt. J. Reine Angew. Math. **177**, 215–237 (1937)

[Has37d] H. Hasse, Über die Riemannsche Vermutung in Funktionenkörpern, in C. R. du Congrès Internat. Math. Oslo 1936, vol. 1, pp. 189–206 (1937)

[Has42a] H. Hasse, Der n-Teilungskörper eines abstrakten elliptischen Funktionenkörpers als Klassenkörper, nebst Anwendung auf den Mordell-Weilschen Endlichkeitssatz. Math. Z. **48**, 48–66 (1942)

[Has42b] H. Hasse, Zur arithmetischen Theorie der algebraischen Funktionenkörper. Jahresber. Dtsch. Math.-Ver. **52**, 1–48 (1942)

[Has43a] H. Hasse, Internationale Mathematikertagung in Rom im November 1942. Jber. Deutsche Math. Ver. **53**, 21–22 (1943). 2.Abteilung

[Has43b] H. Hasse, Punti razionali sopra curve algebriche a congruenze. Reale Accademia d'Italia, Fondazione Alessandro Volta. Atti dei Convegni 9(1939), pp. 85–140 (1943)

[Has45] H. Hasse, Überblick über die neuere Entwicklung der arithmetischen Theorie der algebraischen Funktionen. Atti Convegno Mat. Roma **1942**, 25–33 (1945)

[Has49] H. Hasse, *Zahlentheorie* (Akademie–Verlag, Berlin, 1949) XII, 468 pp.

[Has55] H. Hasse, Zetafunktionen und L-Funktionen zu einem arithmetischen Funktionenkörper vom Fermatschen Typus. Abh. Deutsch. Akad. wiss. Berlin, Math.-Naturw. Kl. **1954**(4), 70 S. (1955)

[Has67] H. Hasse, Vorlesungen über Klassenkörpertheorie. (Physica–Verlag, Würzburg, 1967), 275 pp.

[Has68] H. Hasse, *The Riemann Hypothesis in Algebraic Function Fields over a Finite Constants Field*. Lecture Notes (Pennsylvania State University, State College, Spring 1968)

[Hec87] E. Hecke, Analysis und Zahlentheorie. Vorlesung Hamburg 1920. Bearbeitet von Peter Roquette. Dokumente zur Geschichte der Mathematik. Vieweg, Braunschweig (1987). XXV u. 234 pp.

[Hen37] K. Hensel, Über den Zusammenhang zwischen den Kongruenzgruppen eines algebraischen Körpers für alle Potenzen eines Primteilers als Modul. J. Reine Angew. Math. **177**, 82–93 (1937)

[Her21] G. Herglotz, Zur letzten Eintragung im Gaußschen Tagebuch. Leipz. Ber. **73**, 271–276 (1921). Ges. Schriften, pp. 415–420

[Her79] G. Herglotz, Gesammelte Schriften. Herausgegeben im Auftrage der Akademie der Wissenschaften in Göttingen von Hans Schwerdtfeger. Göttingen: Vandenhoeck & Ruprecht. XL, 652 S. DM 128.00 (1979)

[Hey29] K. Hey, Analytische Zahlentheorie in Systemen hyperkomplexer Zahlen. Dissertation, Hamburg (1929), 49 p.

[Hil97] D. Hilbert, Die Theorie der algebraischen Zahlkörper. Jahresber. Dtsch. Math. Ver., 4:I–XVIII u. 175–546, 1897. Englische Übersetzung: The Theory of Algebraic Number Fields (Springer, Heidelberg, 1998)

[Hil98] D. Hilbert, On the theory of the relative abelian number fields. (Über die Theorie der relativ-Abel'schen Zahlkörper.). Gött. Nachr. **1898**, 370–399 (1898)

[HL02] K. Hensel, G. Landsberg, Theorie der algebraischen Funktionen einer Variablen und ihre Anwendungen auf algebraische Kurven und abelsche Integrale. Teubner, Leipzig (1902) XVI 707 pp.

[Hur86] A. Hurwitz, Über algebraische Korrespondenzen und das verallgemeinerte Korrespondenzprinzip. Berichte v.d. sächsischen Gesellsch. d. Wissenschaften, mathematisch-physikalische Klasse. **1886**, 10–38 (1886). Wiederabdruck in Math. Ann. Bd.28 (1887) 561–585

[HW36] H. Hasse, E. Witt, Zyklische unverzweigte Erweiterungskörper vom Primzahlgrad p über einem algebraischen Funktionenkörper der Charakteristik p. Monatsh. Math. Phys. **43**, 477–492 (1936)

[Jac07] E. Jacobsthal, Über die Darstellung der Primzahlen der Form 4n+1 als Summe zweier Quadrate. J. Reine Angew. Math. **132**, 238–246 (1907)

[Käh64] E. Kähler, Geppert, Harald, in *Neue Deutsche Biographie*, vol. 6, p. 247 (1964)

[Kan80] E. Kani, Eine Verallgemeinerung des Satzes von Castelnuovo-Severi. J. Reine Angew. Math. **318**, 178–220 (1980)

[Kor19] H. Kornblum, Über die Primfunktionen in einer arithmetischen Progression. Math. Z. **5**, 100–111 (1919)

[Kru35] W. Krull, Idealtheorie. Erg. d. Math. u. ihrer Grenzgebiete. Julius (Springer, Berlin, 1935). V, 172 pp.

[Küh02] H. Kühne, Eine Wechselbeziehung zwischen Funktionen mehrerer Unbestimmter, die zu Reziprozitätsgesetzen führt. J. Reine Angew. Math. **124**, 121–133 (1902)

[Küh03] H. Kühne, Angenäherte auflösung von kongruenzen nach primmodulsystemen in zusammenhang mit den einheiten gewisser körper. J. Reine Angew. Math. **126**, 102–115 (1903)

[Kum87] E. Kummer, Zwei neue Beweise der allgemeinen Reciprocitätsgesetze unter den Resten und Nichtresten der Potenzen, deren Grad eine Primzahl ist. J. Reine Angew. Math. **100**, 10–50 (1887). Abhandl. Königl. Akademie d. Wissenschaften zu Berlin (1861), pp. 81–122

[Lam57] E. Lamprecht, Zur Eindeutigkeit von Funktionalprimdivisoren. Arch. Math. **8**, 30–38 (1957)

[Lan18] E. Landau, Einführung in die elementare und analytische Theorie der algebraischen Zahlen und der Ideale. Leipzig: B. G. Teubner. (1918). VII, 143 pp.

[Lan35] W. Landherr, Über einfache Liesche Ringe. Abh. Math. Semin. Univ. Hamburg **11**, 41–64 (1935)

[Lef24] S. Lefschetz, L'analysis situs et la géométrie algébrique. Collection de monographies sur la théorie des fonctions. Paris: Gauthier-Villars, vi, 154 S. 8° (1924)

[Lef28] S. Lefschetz, Correspondences between algebraic curves, in *Selected topics in Algebraic Geometry*, V. Snyder et al. Bulletin of the National Research Council, vol. 63 (1928), pp. 381–448

[Lem00] F. Lemmermeyer, *Reciprocity Laws. From Euler to Eisenstein* (Springer, Berlin, 2000), XIX, 487 pp.

[LR12] F. Lemmermeyer, P. Roquette (eds.), Die mathematischen Tagebücher von Helmut Hasse 1923–1935. With an Introduction in English. Universitäts–Verlag, Göttingen, 2012. 563 pp.

[Lut37] E. Lutz, Sur l'equation $y^2 = x^3 - Ax - B$ dans les corps p-adiques. J. Reine Angew. Math. **177**, 238–247 (1937)

[Mac71] R.E. MacRae, On unique factorization in certain rings of algebraic functions. J. Algebra **17**, 243–261 (1971)

[Man65] Y. Manin, The Hasse-Witt matrix of an algebraic curve. Am. Math. Soc. Transl. II. Ser. **45**, 245–264 (1965). Russian original 1961

[Mil16] J.S. Milne, The Riemann hypothesis over finite fields - from Weil to the present day, in *The Legacy of Bernhard Riemann After One Hundred and Fifty Years*, vol. II (International Press Somerville, MA, 2016), pp. 487–565; (Higher Education Press, Beijing, 2016)

[Mor06] P. Morton, Explicit identities for invariants of elliptic curves. J. Number Theory **120**, 234–271 (2006)

[Mor07] P. Morton, Ogg's theorem via explicit congruences for class equations. (2007, preprint)

[Mor22] L.J. Mordell, On the rational solutions of the indeterminate equations of the third and fourth degrees. Proc. Camb. Philos. Soc. **21**, 179–192 (1922)

[Mor33] L.J. Mordell, The number of solutions of some congruences in two variables. Math. Z. **37**, 193–209 (1933)

[Mor71] L.J. Mordell, Some aspects of Davenport's work. Acta Arith. **18**, 5–11 (1971)

[MQ72] M.L. Madan, C.S. Queen, Algebraic function fields of class number one. Acta Arith. **20**, 423–432 (1972)

[MT58] A. Mattuck, J. Tate, On the inequality of Castelnuovo-Severi. Abh. Math. Semin. Univ. Hamburg **22**, 295–299 (1958)

[Neh33] H. Nehrkorn, Über absolute Idealklassengruppen und Einheiten in algebraischen Zahlkörpern. Abh. Math. Semin. Univ. Hamburg **9**, 318–334 (1933)

[Noe25] E. Noether, Gruppencharaktere und Idealtheorie. Jahresber. Dtsch. Math.-Ver. **34**, 144 (1925). 2. Abteilung

[Ogg75] A.P. Ogg, Automorphisms de courbes modulaires. Sèminaire Delange-Pisot-Poitou (Théorie des nombres), 1975(7) (1975)

[OS16] F. Oort, N. Schappacher, Early history of the Riemann hypothesis in positive characteristic, in *The Legacy of Bernhard Riemann After One Hundred and Fifty Years*, vol. II (International Press, Somerville, MA, 2016), pp. 595–631; (Higher Education Press, Beijing, 2016)

[Rei76] C. Reid, *Courant in Göttingen and New York.* (Springer, New York, 1976), 314 pp.

[Roh64] H. Rohrbach, Helmut Hasse und das Crellesche journal. J. Reine Angew. Math. **214–215**, 443–444 (1964)

[Roq05] P. Roquette, *The Brauer-Hasse-Noether Theorem in Historical Perspective.*. Schriftenreihe der Heidelberger Akademie der Wissenschaften, vol. 15 (Springer, Berlin, 2005). I, 77 pp.

[Roq53] P. Roquette, Arithmetischer Beweis der Riemannschen Vermutung in Kongruenzfunktionenkörpern beliebigen Geschlechts. J. Reine Angew. Math. **191**, 199–252 (1953)

[Roq58] P. Roquette, Über den Riemann-Rochschen Satz in Funktionenkörpern vom Transzendenzgrad 1. (On the Riemann-Roch theorem in function fields of transcendence degree 1). Math. Nachr. **19**, 375–404 (1958)

[Roq76] P. Roquette, On the division fields of an algebraic function field of one variable. An estimate for their degree of irrationality. Houston J. Math. **2**, 251–287 (1976)

[Ros52] M. Rosenlicht, Equivalence relations on algebraic curves. Ann. Math. (2) **56**, 169–191 (1952)

[Sch25] F.K. Schmidt, Allgemeine Körper im Gebiet der höheren Kongruenzen. PhD thesis, Universität Freiburg (1925)

[Sch26] O. Schreier, Über eine Arbeit von Herrn Tschebotareff. Abh. Math. Semin. Univ. Hamburg **5**, 1–6 (1926)

[Sch31a] F.K. Schmidt, Analytische Zahlentheorie in Körpern der Charakteristik p. Math. Z. **33**, 1–32 (1931)

[Sch31b] F.K. Schmidt, Die Theorie der Klassenkörper über einem Körper algebraischer Funktionen in einer Unbestimmten und mit endlichem Koeffizientenbereich. Sitzungsberichte Erlangen **62**, 267–284 (1931)

[Sch35a] O.F.G Schilling, Über gewisse Beziehungen zwischen der Arithmetik hyperkompler Zahlsysteme und algebraischer Zahlkörper. Math. Ann. **111**, 372–398 (1935)

[Sch35b] H.L. Schmid, Zyklische algebraische Funktionenkörper vom Grade p^n über endlichem Konstantenkörper der Charakteristik p. J. Reine Angew. Math. **175**, 108–123 (1935)

[Sch36] F.K. Schmidt, Zur arithmetischen Theorie der algebraischen Funktionen. I. Beweis des Riemann-Rochschen Satzes für algebraische Funktionen mit beliebigem Konstantenkörper. Math. Zeitschr. **41**, 415–438 (1936)

[Sch39] F.K. Schmidt, Zur arithmetischen Theorie der algebraischen Funktionen. II: Allgemeine Theorie der Weierstraßpunkte. Math. Z. **45**, 75–92 (1939)

[Sch41] H.L. Schmid, Kongruenzzetafunktionen in zyklischen Körpern. Abh. Preuß. Akad. Wiss., math.-naturw. Kl. **1941**(14), 30 S, (1941)

[Sch50] O.F.G. Schilling, *The Theory of Valuations*. Mathematical Surveys, vol. IV (American Mathematical Society, Providence, RI, 1950), 253 pp.

[Sch87] N. Schappacher, Das mathematische Institut der Universität Göttingen 1929–1950, in *Die Universität Göttingen unter dem Nationalsozialismus*, ed. by H. Becker, andere, pp. 345–373. K. G. Saur (1987)

[Sch97] N. Schappacher, Some milestones of lemniscatomy, in *Algebraic Geometry. Proceedings of the Bilkent Summer School*, Ankara, August 7–19, 1995 (Marcel Dekker, New York, NY, 1997), pp. 257–290

[Sch07] N. Schappacher, A historical sketch of B. L. van der Waerden's work in algebraic geometry: 1926–1946. In Episodes in the history of modern algebra (1800–1950). Based on a workshop held at MSRI, Berkeley, CA, USA, April 2003 (American Mathematical Society, Providence, RI, 2007); (London Mathematical Society, London 2007), pp. 245–283

[Sch14] O. Schreier, Briefe an Karl Menger und Helmut Hasse. Herausgegeben und kommentiert von Alexander Odefey., Algorismus. Studien zur Geschichte der Mathematik und der Naturwissenschaften, vol. 81 Dr. Erwin Rauner Verlag, Augsburg (2014), 149 pp.

[Sch27] F.K. Schmidt, Zur Zahlentheorie in Körpern von der Charakteristik *p*. Sitzungsberichte Erlangen **58/59**, 159–172 (1926/1927)

[Sen25] P. Sengenhorst, Über Körper der Charakteristik *p*. Math. Zeitschr. **24**, 1–39 (1925)

[Sev26] F. Severi, Trattato di geometria algebrica. Vol.I. Parte I. Geometria delle serie lineari. Zanichelli, Bologna, 1926. 358 pp.

[Sha51] I. Shafarevic, Das allgemeine Reziprozitätsgesetz. Mat. Sbornik, n.S. **26**(68), 113–146 (1951). Russisch

[Sha57] I. Shafarevic, Exponents of elliptic curves. Dokl. Akad. Nauk SSSR, n.S. **114**, 714–716 (1957). Russisch

[Shi55] G. Shimura, Reduction of algebraic varieties with respect to a discrete valuation of the basic field. Am. J. Math. **77**, 134–176 (1955)

[Shi67] K. Shiratani, Über singuläre Invarianten elliptischer Funktionenkörper. J. Reine Angew. Math. **226**, 108–115 (1967)

[Sie29] C.L. Siegel, Über einige Anwendungen diophantischer Approximationen. Abhandlungen Akad. Berlin **1929**(1), 70pp. (1929)

[Söh35] H. Söhngen, Zur komplexen Multiplikation. Math. Ann. **111**, 302–328 (1935)

[SS92] N. Schappacher, E. Scholz (eds.), Oswald Teichmüller – Leben und Werk. Jahresber. Dtsch. Math.-Ver. **94**, 1–39 (1992)

[Ste10] E. Steinitz, Algebraische Theorie der Körper. J. Reine Angew. Math. **137**, 167–309 (1910)

[Ste69] S.A. Stepanov, Über die Anzahl der Punkte einer hyperelliptischen Kurve über einem einfachen endlichen Körper. Izv. Akad. Nauk SSSR, Ser. Mat. **33**, 1171–1181 (1969)

[Sti09] H. Stichtenoth, *Algebraic Function Fields and Codes*, 2nd edn. (Springer, Berlin, 2009). XII+355 pp.

[Sti90] L. Stickelberger, Über eine Verallgemeinerung der Kreistheilung. Math. Annalen. **37**, 321–367 (1890)

[Tei36] O. Teichmüller, Differentialrechnung in Charakteristik *p*. J. Reine Angew. Math. **175**, 89–99 (1936)

[Tob03] R. Tobies, Briefe Emmy Noethers an P.S. Alexandroff. NTM N.S. **11**(2), 100–115 (2003)

[Tse33] C. Tsen, Divisionsalgebren über Funktionenkörpern. Nachr. Ges. Wiss. Göttingen, Math.–Phys. Kl. I, **1933**(44), 335–339 (1933)

[Ul100] P. Ullrich, Emil Artins unveröffentlichte Verallgemeinerung seiner Dissertation. Mitt. Math. Ges. Hamburg **19**, 173–194 (2000)

[van47] B.L. van der Waerden, Divisorenklassen in algebraischen Funktionenkörpern. Comment. Math. Helv. **20**, 68–80 (1947)

[vdW67] B.L. van der Waerden, Algebra. Teil II. Unter Benutzung von Vorlesungen von E. Artin und E. Noether. Fünfte Auflage. Heidelberger Taschenbücher, Band 23 (Springer, New York, 1967)

[vdW83] B.L. van der Waerden, Zur algebraischen Geometrie. Selected Papers. (Springer, Berlin, 1983), VII + 479 pp.

[Web08] H. Weber, Lehrbuch der Algebra. Dritter Band: Elliptische Funktionen und algebraische Zahlen., volume 3. Friedr. Vieweg u. Sohn, Braunschweig, 1908. XVI, 733 pp.

[Wei29] A. Weil, L'arithmétique sur les courbes algébriques. Acta Math. **52**, 281–315 (1929)

[Wei37] J. Weissinger, Theorie der Divisorenkongruenzen. Abh. math. Sem. Univ. Hamburg **12**, 115–126 (1937)

[Wei38a] A. Weil, Généralisation des fonctions abéliennes. J. Math. Pures Appl. **9**(17), 47–87 (1938)

[Wei38b] A. Weil, Zur algebraischen Theorie der algebraischen Funktionen. J. Reine Angew. Math. **179**, 129–133 (1938)

[Wei40] A. Weil, Sur les fonctions algébriques à corps de constantes fini. C. R. Acad. Sci. Paris **210**, 592–594 (1940)

[Wei41] A. Weil, On the Riemann hypothesis in function fields. Proc. Natl. Acad. Sci. USA **27**, 345–347 (1941)

[Wei46] A. Weil, *Foundations of Algebraic Geometry*. Colloquium Publications, vol. XXIX (American Mathematical Society, Providence, RI, 1946)

[Wei48a] A. Weil, Sur les courbes algébriques et les variétés qui s'en deduisent. Actualités scientifiques et industrielles, vol. 1048 (Hermann & Cie, Paris, 1948), 85 pp.

[Wei48b] A. Weil, Variétés abéliennes et courbes algébriques, Actualités scientifiques et industrielles, vol. 1064 (Hermann & Cie, Paris, 1948), 163 pp.

[Wei52] A. Weil, Jacobi sums as "Größencharaktere". Trans. Am. Math. Soc. **73**, 487–495 (1952)

[Wei56] A. Weil, Abstract versus classical algebraic geometry, in *Proceedings of International Congress in Mathematics*, Amsterdam 3, pp. 550–558 (1956)

[Wei59] A. Weil, Adèles et group algébriques. In Sémin. Bourbaki 11 (1958/1959), Exp. No.186 (1959). 9pp.

[Wei79a] A. Weil, Œuvres scientifiques. Collected papers. Vol. I (1926–1951). (Springer, Berlin, 1979)

[Wei79b] A. Weil, Review: the collected papers of Emil Artin, in *Collected Papers, Vol. III (1964–1978)*, ed, by A. Weil, (1979), pp. 237–238

[Wei80] A Weil, Lettre à Artin (1942), in *A. Weil, Collected Papers.*, vol. 1 (Springer, Berlin, 1980), pp. 280–298

[Wei91] A. Weil, *Souvenirs d'apprentissage* (Birkhäuser, Basel, 1991)

[Wit34a] E. Witt, Riemann–Rochscher Satz und ζ–Funktion im Hyperkomplexen. Math. Ann. **110**, 12–28 (1934)

[Wit34b] E. Witt, Über ein Gegenbeispiel zum Normensatz. Math. Z. **39**, 462–467 (1934)

[Wit36] E. Witt, Zyklische Körper und Algebren der Charakteristik p vom Grad p^n. Struktur diskret bewerteter perfekter Körper mit vollkommenem Restklassenkörper der Charakteristik p. J. Reine Angew. Math. **176**, 126–140 (1936)

[Wuß08] H. Wußing, Zur Emigration von Emil Artin. in *Mathematics celestial and terrestrial. Festschrift für Menso Folkerts zum 65. Geburtstag*, ed. by J.W. Dauben, S. Kirschner, A. Kühne. Acta Historica Leopoldina, vol. 54, pp. 705–716 (2008)

[WZ91] B. Weis, H.G. Zimmer, Artins Theorie der quadratischen Kongruenzfunktionenkörper und ihre Anwendung auf die Berechnung der Einheiten- und Klassengruppen. (Artin's theory of quadratic congruence function fields and their application to the calculation of unit und class groups). Mitt. Math. Ges. Hamburg **12**(2), 261–286 (1991)

[Zas34] H. Zassenhaus, Zum Satz von Jordan-Hölder-Schreier. Abh. Math. Semin. Univ. Hamburg **10**, 106–108 (1934)

[Zas37] H. Zassenhaus, Lehrbuch der Gruppentheorie. Bd. 1. Hamburg. Math. Einzelschriften, vol. 21 (B.G. Teubner, Leipzig, Berlin, 1937), VI, 152 pp.

[Zor33] M. Zorn, Note zur analytischen hyperkomplexen Zahlentheorie. Abh. Math. Semin. Univ. Hamburg **9**, 197–201 (1933)

Index

Adele, 191
Ahlfors, 204
Albert, A. A., 152
Alexandroff, 40
Artin, 3, 13–31, 56, 131, 133, 181, 183, 212, 213
Artin-Schreier equation, 74
Artin's criterion, 27, 52
Audin, 201, 204

Baer, 83
Base field, 9
    extension of, 26, 46
Behrbohm, 99, 154
Berzolari, 185
Bieberbach, 187
Blaschke, 31, 91, 153, 154, 182, 183, 186, 188
Bol, 154
Bombieri, 7, 216
Bompiani, 186
Brandt, 90
Brauer, Richard, 155, 163

Caratheodory, 186, 188
Cartan, Elie, 200, 205
Cartan, Henri, 201, 204, 205, 208
Castelnuovo, 211
Castelnuovo-Severi
    inequality of, 166, 184, 212
Chabauty, 200
Chebotarev, 41
Chern, 90

Chevalley, 90, 190, 192, 193, 196, 197, 199, 200, 208
Chow, 90
Coarse equivalence, 145
Complex multiplication, 27
    algebraic, 126
    classic, 85
Correspondences, 143
Courant, 2, 5, 140, 141, 182

Déligne, 215
Davenport, 4, 5, 36, 60–78, 81–101, 105, 135, 194, 196, 197
Davenport-Hasse function field, 73
Dedekind, 12, 16, 19, 42
Deuring, 6, 36, 85, 104, 117–128, 137–146, 148–151, 153, 155, 186, 188, 195, 205, 207, 208, 210, 211, 216
Dieudonné, 209, 215
Different
    of separable extension, 45
    of transcendental divisors, 159
Differential, 107, 191
Discriminant, 170
    estimate of, 170
Divisor, 10
    degree of, 10
    integer, 11
    non-special, 169
    principal, 10
Divisor class, 10
    canonical, 44
    coarse, 145

© Springer Nature Switzerland AG 2018
P. Roquette, *The Riemann Hypothesis in Characteristic* p
in Historical Perspective, Lecture Notes in Mathematics 2222,
https://doi.org/10.1007/978-3-319-99067-5

dimension of, 44
Double field, 96, 143, 158, 186

Ehresmann, 201
Eichler, 120, 193
Exponential sum, 76

Fermat generalized function field, 69
Fraenkel, 56
Frei, 15, 19
Freitag, 215
Frobenius operator, 84, 102, 175
Function field, 9
    elliptic, 14, 81, 83
        supersingular, 118
    hyperelliptic, 14
    quadratic, 14
Furtwängler, 13

Gauss, 32
Gauss sum, 71
Genus, 44
Geppert, 153, 154, 184–187
Good reduction, 124
Grothendieck, 215

Hadamard, 203
Hall, Philip, 151
Hasse, 4–212, 215, 216
Hasse invariant A, 107
Hasse-Witt matrix, 111
Hecke, 5, 31, 56, 91, 182, 192
Hensel, 5, 56, 115, 192, 196
Herglotz, 3, 4, 14, 15, 31, 32, 37
Hey, 90
Higher derivation, 115
Hilbert, 2, 5, 31, 37, 75, 132, 133, 157, 159,
      160, 176
Humbert, 194
Hurwitz, 150, 205, 207

Idele, 192
Intersection multiplicity, 178
Iyanaga, 65

Jacobi sum, 71
Jacobi symbol, 20
Jacobian, 93, 112

Jacobsthal, 29
Jarden, 56, 216
Julia, 190, 199, 201, 208, 210
Jung, 153, 154

Kähler, 91
Kiehl, 215
Klein, 2
Kornblum, 15, 23
Köthe, 139
Krull, 40, 42, 154
Kühne, 19

Lamprecht, 125
Landau, 2, 23
Landherr, 91
Landsberg, 115
Lefschetz, 150, 151, 182, 208
Lemmermeyer, 19, 33
Loewy, 40
L-polynomial, 47
L-series, 22, 72

Maak, 154
MacRae, 25
Madan, 25
Mahler, 83
Mattuck, 215
Menger, 41
Meromorphism, 94
Methodenrein, 127
Mordell, 60–67, 69, 75, 82, 131
Morton, 120
Multiplier, 86

Nehrkorn, 90
Neron, 215
Neumann, Hanna, 154
Nevanlinna, 136, 204
Noether, Emmy, 2, 17, 40–42, 73, 90, 137, 139,
      151, 153, 155, 182

Petersson, 90, 154, 195
Pisot, 200
Prime divisor, 10
    transcendental, 97, 158

Queen, 25

Reciprocity law
    Artin's, 73
    power, 19
    quadratic, 19
Redei, 99
Reich, 182
Reid, Constance, 26
Reidemeister, 186
RHp, 12, 173
Riemann-Hurwitz formula, 45, 168
Riemann-Roch Theorem, 44, 191
Roger, 201
Rohrbach, 154
Rosati, 150, 185
    anti-automorphism, 104, 207
Runge, 36

Schappacher, 35, 152
Schertz, 89
Schilling, O.F.G., 150, 151, 187, 209
Schmid, H.L., 40, 147, 154, 194, 209
Schmidt, Erhard, 186
Schmidt, F.K., 5, 19, 37, 40–51, 140, 191, 196,
    197
Schneider, Theodor, 36
Schreier, 41
Sengenhorst, 41
Separating element, 40
Severi, 150, 183–186, 200, 205, 207, 210, 211,
    215
Shafarevich, 147
Shimura, 125, 127
Siegel, 139, 176, 194, 198, 208, 209
Söhngen, 91
Stickelberger, 75
Supersingular, 118
Süss, 186

Szegö, 36

Takagi, 133
Tate, 215
Teichmüller, 99, 137, 138, 154
Toeplitz, 56, 135
Tornier, 137, 140
Tsen, 73

Ullrich, 14, 31

van der Waerden, B.L., 43, 90, 136, 137,
    152–154, 209

Weber, 16
Weierstrass, 35
Weil, 6, 90, 127, 132–134, 152, 154, 155,
    188–200, 204–212, 215, 216
Weil conjectures, 215
Weissinger, 193
Weyl, 151
Wirtinger, 186
Witt, 6, 46, 73, 94, 95, 99, 154, 193, 195, 196
Witt vector, 147
Wußing, 182

Zariski, 152
Zassenhaus, 91, 154
Zeta function, 11, 43
    Artin's, 20
    functional equation, 49
    of higher dimensional variety, 194
Zorn, 90

LECTURE NOTES IN MATHEMATICS  Springer

Editors in Chief: J.-M. Morel, B. Teissier;

Editorial Policy

1. Lecture Notes aim to report new developments in all areas of mathematics and their applications – quickly, informally and at a high level. Mathematical texts analysing new developments in modelling and numerical simulation are welcome.

   Manuscripts should be reasonably self-contained and rounded off. Thus they may, and often will, present not only results of the author but also related work by other people. They may be based on specialised lecture courses. Furthermore, the manuscripts should provide sufficient motivation, examples and applications. This clearly distinguishes Lecture Notes from journal articles or technical reports which normally are very concise. Articles intended for a journal but too long to be accepted by most journals, usually do not have this "lecture notes" character. For similar reasons it is unusual for doctoral theses to be accepted for the Lecture Notes series, though habilitation theses may be appropriate.

2. Besides monographs, multi-author manuscripts resulting from SUMMER SCHOOLS or similar INTENSIVE COURSES are welcome, provided their objective was held to present an active mathematical topic to an audience at the beginning or intermediate graduate level (a list of participants should be provided).

   The resulting manuscript should not be just a collection of course notes, but should require advance planning and coordination among the main lecturers. The subject matter should dictate the structure of the book. This structure should be motivated and explained in a scientific introduction, and the notation, references, index and formulation of results should be, if possible, unified by the editors. Each contribution should have an abstract and an introduction referring to the other contributions. In other words, more preparatory work must go into a multi-authored volume than simply assembling a disparate collection of papers, communicated at the event.

3. Manuscripts should be submitted either online at www.editorialmanager.com/lnm to Springer's mathematics editorial in Heidelberg, or electronically to one of the series editors. Authors should be aware that incomplete or insufficiently close-to-final manuscripts almost always result in longer refereeing times and nevertheless unclear referees' recommendations, making further refereeing of a final draft necessary. The strict minimum amount of material that will be considered should include a detailed outline describing the planned contents of each chapter, a bibliography and several sample chapters. Parallel submission of a manuscript to another publisher while under consideration for LNM is not acceptable and can lead to rejection.

4. In general, **monographs** will be sent out to at least 2 external referees for evaluation.

   A final decision to publish can be made only on the basis of the complete manuscript, however a refereeing process leading to a preliminary decision can be based on a pre-final or incomplete manuscript.

   Volume Editors of **multi-author works** are expected to arrange for the refereeing, to the usual scientific standards, of the individual contributions. If the resulting reports can be forwarded to the LNM Editorial Board, this is very helpful. If no reports are forwarded or if other questions remain unclear in respect of homogeneity etc, the series editors may wish to consult external referees for an overall evaluation of the volume.

5. Manuscripts should in general be submitted in English. Final manuscripts should contain at least 100 pages of mathematical text and should always include

   – a table of contents;
   – an informative introduction, with adequate motivation and perhaps some historical remarks: it should be accessible to a reader not intimately familiar with the topic treated;
   – a subject index: as a rule this is genuinely helpful for the reader.
   – For evaluation purposes, manuscripts should be submitted as pdf files.

6. Careful preparation of the manuscripts will help keep production time short besides ensuring satisfactory appearance of the finished book in print and online. After acceptance of the manuscript authors will be asked to prepare the final LaTeX source files (see LaTeX templates online: https://www.springer.com/gb/authors-editors/book-authors-editors/manuscriptpreparation/5636) plus the corresponding pdf- or zipped ps-file. The LaTeX source files are essential for producing the full-text online version of the book, see http://link.springer.com/bookseries/304 for the existing online volumes of LNM). The technical production of a Lecture Notes volume takes approximately 12 weeks. Additional instructions, if necessary, are available on request from lnm@springer.com.

7. Authors receive a total of 30 free copies of their volume and free access to their book on SpringerLink, but no royalties. They are entitled to a discount of 33.3 % on the price of Springer books purchased for their personal use, if ordering directly from Springer.

8. Commitment to publish is made by a *Publishing Agreement*; contributing authors of multiauthor books are requested to sign a *Consent to Publish form*. Springer-Verlag registers the copyright for each volume. Authors are free to reuse material contained in their LNM volumes in later publications: a brief written (or e-mail) request for formal permission is sufficient.

Addresses:
Professor Jean-Michel Morel, CMLA, École Normale Supérieure de Cachan, France
E-mail: moreljeanmichel@gmail.com

Professor Bernard Teissier, Equipe Géométrie et Dynamique,
Institut de Mathématiques de Jussieu – Paris Rive Gauche, Paris, France
E-mail: bernard.teissier@imj-prg.fr

Springer: Ute McCrory, Mathematics, Heidelberg, Germany,
E-mail: lnm@springer.com

Printed in the United States
By Bookmasters